Lecture Notes in Mathematics

Edited by A. Dold and B. Eckmann

801

Klaus Floret

Weakly Compact Sets

Lectures Held at S.U.N.Y., Buffalo, in Spring 1978

Springer-Verlag
Berlin Heidelberg New York 1980

Author

Klaus Floret
Mathematisches Seminar der Universität Kiel
Olshausenstr. 40–60
2300 Kiel
Federal Republic of Germany

AMS Subject Classifications (1980):
Primary: 46A05, 46A50
Secondary: 41A65, 46A25, 46B10, 46E15, 46E30, 54C35, 54D30, 54D60

ISBN 3-540-09991-3 Springer-Verlag Berlin Heidelberg New York
ISBN 0-387-09991-3 Springer-Verlag New York Heidelberg Berlin

Printed in Germany

Printing and binding: Beltz Offsetdruck, Hemsbach/Bergstr.
2141/3140-543210

Weakly compact subsets, i.e. sets which are compact with respect to the weak topology of a Banach-space or more generally: of a locally convex space play an important rôle in many questions of analysis. Among them are characterizations of reflexivity, characterizations of subsets with elements of least distance in linear and convex approximation theory, ranges of vector measures and existence theorems in optimal control theory, pointwise convergence of sequences of functions, minimax-theorems, separation properties of convex sets.

The intention of these lecture notes is to prove the main results on weak compactness due to W.F. Eberlein, V.L. Šmulian, M. Krein, A. Grothendieck, and R.C. James as well as to go into some of the questions mentioned above. There are three foci: the theorems on countable compactness, on sequential compactness, and the supremum of linear functionals. The linking element is A. Grothendieck's interchangeable double-limit property. The results on countable and sequential compactness are, as usual, first proved in spaces of continuous functions, equipped with the topology of pointwise convergence. The approach to R.C. James' theorem and its various applications is the original one in the form which was given by J.D. Pryce: His proof is just checked carefully and the result stated as a double-limit-theorem which implies many of the applications of other versions due to S. Simons and M. DeWilde. A short look into the contents shows that emphasis is put on R.C. James' theorem . A reader who is just interested in this, may start with §5 provided she or he accepts the W.F. Eberlein-A. Grothendieck-theorem (1.6.), the W.F. Eberlein-V.L. Šmulian-theorem (3.10.) and a consequence of it, A. Grothendieck's theorem on weak compactness in C(K) (4.2. and 4.4.).

The typical reader whom I have in mind knows the basic facts on locally convex spaces and became somehow interested in weakly compact sets: either by some applications, or by their rôle in the general theory, some properties of them, or from a topological point of view. Consequently, the introductory remarks on locally convex spaces do not serve the purpose of explaining what locally convex spaces are and what they are for; they simply try to go through those parts of the theory which will be needed later on - with the additional benefit that some more or less standard notation will be fixed. Relative to these facts and some knowledge of topology (e.g. a compact space is a Hausdorff topological space such that every net has a cluster-point) the exposition is self-contained.

There are exercises attached to each section: I simply believe that it is much easier to understand a result once one has solved a related problem. At first glance, some of the exercises may seem to be difficult, but on the basis of the foregoing text and the hints the reader should be able to master them. I do not claim that a result stated as an exercise is easy in an "absolute" sense: I only say that at this point of the text there is enough information available to prove it without too much extra effort.

The notes are based on lectures I gave at the State University of New York at Buffalo during the Spring semester of 1978. They may serve as well as a basis for a seminar.

While preparing these lectures I was deeply influenced by the seminar-notes [7] of M. De Wilde and two papers of J.D. Pryce, one [40] presenting a smoothened proof of R.C. James' theorem, the other one [41] dealing with H.D. Fremlin's notion of an "angelic" space.

I thank the Department of Mathematics of S.U.N.Y.A.B. for the invitation to spend the academic year 1977/78 in Buffalo and the colleagues and friends there who created a kind and open atmosphere for me. P. Dierolf, W. Govaerts, M. Wriedt, and V. Wrobel made many valuable remarks on the text. Special thanks to Mrs. Marie Daniel who typed the manuscript with great patience and diligence. I am grateful to the editors for accepting these notes for publication in the Lecture Notes Series.

October 1978 Klaus Floret

CONTENTS

§0 SOME FUNDAMENTALS OF LOCALLY CONVEX SPACES

0.1. A locally convex space (E,τ) is a topological vector space (here: always real and Hausdorff) whose topology τ has a basis $\mathfrak{U}_E(0)$ of absolutely convex (= convex and symmetric) neighbourhoods of zero. Equivalently, the topology can be given by a family $\mathcal{P}$ of seminorms, which is filtrating (for all $p,q \in \mathcal{P}$ there is a $r \in \mathcal{P}$ with $\max(p,q) \leq r$) and separating. The connection is given by the Minkowski-or gauge-functional

$$m_A(x) := \inf \{\lambda > 0 \mid x \in \lambda A\}$$

of an absolutely convex set $A \subset E$, which is finite on the linear hull span A of A.

0.2. The fundamental tool in Functional Analysis is the subspace $(E,\tau)' = E'$ of the algebraic dual E^* consisting of continuous linear functionals on the locally convex space (E,τ). Together with this (topological) dual, E forms a dual pair of vector-spaces

$$\langle E',E\rangle \quad \text{and} \quad \langle \varphi,x\rangle := \varphi(x) \qquad \varphi \in E',\ x \in E,$$

since $\langle \cdot,x\rangle = o$ implies $x = o$ by the Hahn-Banach theorem. Whenever $\langle G,H\rangle$ is a (separating) dual system of vector-spaces, the product topology on $\mathbb{R}^H$ induces by

$$G \hookrightarrow \mathbb{R}^H$$
$$g \rightsquigarrow (\langle g,h\rangle)_{h \in H}$$

a (separated) locally convex topology on G called the weak topology $\sigma(G,H)$ on G; the dual of G with this topology is just H: $(G,\sigma(G,H))' = H$. For a locally convex space (E,τ) the weak topology of E, that is $\sigma(E,E')$, is coarser than τ. The Tychonoff-theorem on products of compact sets, implies for a zero-neighbourhood $U \in \mathfrak{U}_E(0)$, that the (absolute) polar

$$U^{\circ} := \{\varphi \in E' \mid |\langle \varphi,x\rangle| \leq 1 \quad \text{for all} \quad x \in U\}$$

of U is $\sigma(E',E)$-compact (THEOREM of ALAOGLU-BOURBAKI). The notion of a polar is naturally defined for both members of a dual pairing $\langle G,H\rangle$. The Hahn-Banach-separation-theorem implies that the bipolar of an $A \subset G$ is the closure of the absolute convex hull of A

$$A^{\circ\circ} := (A^{\circ})^{\circ} = \overline{\Gamma A}^{\sigma(G,H)}$$

(BIPOLAR-THEOREM)

0.3. Every equicontinuous set in E' is surely contained in some U^{o}, $U \in \mathfrak{U}_{E}(o)$. The relation

$$m_{U}(x) = \sup_{\varphi \in U^{o}} |\langle \varphi, x \rangle|$$

is just another formulation of the bipolar theorem, hence the topology τ is the topology of uniform convergence on certain absolutely convex $\sigma(E',E)$-compact subsets. Therefore τ is coarser than the Mackey-topology $\mu(E,E')$ of uniform convergence on all absolutely convex $\sigma(E',E)$-compact subsets of E'. The theorem on bipolars, applied to the pairing $\langle E,E^{*}\rangle$ shows that $(E,\mu(E,E'))' = E'$. Since the Mackey-topology again depends only on the pairing, this can be formulated to: If $\langle G,H\rangle$ is a dual pair, a locally convex topology τ has the dual H (i.e. $(G,\tau)' = H$ in G^{*}) if and only if

$$\sigma(G,H) \subset \tau \subset \mu(G,H)$$

(MACKEY-THEOREM on dual pairs). In this case, τ is called compatible with the dual system $\langle G,H\rangle$. In particular, a locally convex topology on G is compatible with $\langle G,H\rangle$ if and only if it is the topology of uniform convergence on sets of a covering of H consisting of $\sigma(H,G)$-compact absolutely convex sets. Though the closed sets differ for different topologies - the Hahn-Banach-theorem implies that the closed convex sets are the same for all compatible topologies of a dual pairing.

0.4. A τ-barrel in a locally convex space E is an absolutely convex closed subset which absorbs all points. Closed, absolutely convex zero-neighbourhoods are certainly barrels and E is called barrelled if every barrel is a zero-neighbourhood. Banach-spaces and Fréchet-spaces (= locally convex, metrizable and complete) are barrelled by the Baire category-theorem.

$B \subset E$ is called bounded if it is absorbed by all $U \in \mathfrak{U}_{E}(0)$. The closure $\overline{B}$, the convex hull $\operatorname{co} B$ and the absolute convex hull ΓB, of a bounded set are bounded. For a bounded, absolutely convex set B the gauge-functional m_{B} is a norm on span B:

$$[\![B]\!] := (\operatorname{span} B, m_{B})$$

is a normed space, (with closed unit ball B, if B is sequentially closed in E) which is continuously embedded in E. In particular, there is a bounded open set if and only if the space is normable (KOLMOGOROFF-CRITERION). A bounded, absolutely convex set B is called a Banach-disc if $[\![B]\!]$ is a Banach-space. This is the case for example if B is sequentially complete. Continuous images and finite sums of Banach-discs are Banach-discs. Since $[\![B]\!]$ is barrelled, a Banach-disc in (E,τ) is absorbed by every τ-barrel (BARREL-LEMMA).

A set $B \subset E$ is weakly (i.e. with respect to $\sigma(E,E')$) bounded if and only if the polar B^o is a $\sigma(E',E)$-barrel; the bipolar theorem and the barrel-lemma applied to the Banach-discs U^o, $U \in \mathfrak{U}_E(o)$, in $\sigma(E',E)$ give, that a set is bounded if and only if it is weakly bounded (MACKEY-THEOREM for bounded sets). In other words: For a dual system $\langle G,H\rangle$ not only the closed convex sets, but also the bounded sets are the same for all compatible topologies.

0.5. For a Banach-space E the norm-topology is the Mackey-topology (this is true e.g. for all barrelled spaces), however, the norm-topology of E' might not even be compatible with $\langle E',E\rangle$. The norm-topology is a special case of the following: For a locally convex space E the <u>strong topology</u> $\beta(E',E)$ on E' is the topology of uniform convergence on all bounded (= $\sigma(E,E')$-bounded) subsets of E (depends again only on the dual-system $\langle E',E\rangle$). Hence the bidual

$$E'' := (E',\beta(E',E))'$$

can be larger than E (as subspaces of E'^*). E is called <u>semi-reflexive</u> if $E = E''$ and <u>reflexive</u> if additionally $\beta(E'',E')$ induces the original topology on E. For a bounded subset $A \subset E$, the polar $A^o \subset E'$ is a typical $\beta(E',E)$-zero-neighbourhood, hence the bipolar A^{oo} (in E'') is a $\sigma(E'',E')$-compact subset of E'' by the Alaoglu-Bourbaki-theorem. Since certainly $\sigma(E'',E')$ induces $\sigma(E,E')$ on E, this means, by the theorem on bipolars, that a locally convex space E is semi-reflexive if and only if every absolutely convex closed bounded subset of E is $\sigma(E,E')$-compact. Surely, in this case all bounded sets are $\sigma(E,E')$-relatively compact. This semi-reflexivity-criterion will be used frequently, mostly without any further reference. A space is reflexive if and only if it is semi-reflexive and barrelled.

0.6. Since by the Alaoglu-Bourbaki-theorem a bounded weakly closed set is always $\sigma(E'',E')$-relatively compact, it is $\sigma(E,E')$-compact if and only if it is $\sigma(E,E')$-complete. Hence a space is semi-reflexive if and only if all absolutely convex closed bounded sets are weakly complete.

If τ_1 and τ_2 are locally convex topologies on a vector space E such that τ_1 is finer than τ_2, i.e.

$$(E,\tau_1) \hookrightarrow (E,\tau_2)$$

is continuous, and τ_1 has a basis of τ_2-closed neighbourhoods of zero then every τ_2-complete subset of E is τ_1-complete. In particular, if τ is any compatible topology on G of a dual pair $\langle G,H\rangle$ and $A \subset G$ is τ-complete, then A is complete for all finer compatible topologies, e.g., $\mu(G,H)$-complete. Weakly complete sets are complete for all compatible topologies, hence semi-reflexive spaces are <u>quasicomplete</u> (= bounded closed sets are complete).

It is important for the investigation of weakly compact sets to have a good description of complete spaces. First note, that every locally convex space E has a completion $\tilde{E}$ (unique up to isomorphy); the sets

$$\overline{U}^{\tilde{E}} =: \tilde{U} \qquad U \in \mathfrak{U}_E(o)$$

form a basis of neighbourhoods of $\tilde{E}$ and $\tilde{E}' = E'$.

THEOREM (A. Grothendieck): Let $\langle G,H\rangle$ be a dual system, $\mathfrak{S}$ a covering of H by absolutely convex, $\sigma(H,G)$-compact subsets,

$$\hat{G} := \{\varphi \in H^* \mid \text{for all } A \in \mathfrak{S}\colon\ \varphi|_A\ \sigma(H,G)\text{-continuous}\} \supset G .$$

Then

(a) $\langle \hat{G},H\rangle$ form a dual system.

(b) All $A \in \mathfrak{S}$ are $\sigma(H,\hat{G})$-compact.

(c) The topology $\tau_{\mathfrak{S}}$ of uniform convergence on all $A \in \mathfrak{S}$ is well-defined on $\hat{G}$ and $\hat{G}$ is $\tau_{\mathfrak{S}}$-complete.

(d) $G \subset \hat{G}$ is $\tau_{\mathfrak{S}}$-dense.

In particular $(\hat{G},\tau_{\mathfrak{S}})$ is a completion of $(G,\tau_{\mathfrak{S}})$.

Proof:

(a) Since $G \subset \hat{G} \subset H^*$ also $\langle \hat{G},H\rangle$ is a dual pair.

(b) Weak topologies are initial topologies, consequently, for $A \in \mathfrak{S}$, the identity map

$$\begin{array}{ccc} (A,\sigma(H,G)) & \hookrightarrow & (A,\sigma(H,\hat{G})) \\ & \overset{\bar{\varphi}}{\searrow} & \downarrow \varphi \in \hat{G} \\ & & \mathbb{R} \end{array}$$

is continuous (since all $\bar{\varphi}$, $\varphi \in \hat{G}$ are continuous by definition) and therefore a homeomorphism.

(c) For $\varphi \in \hat{G}$ and $A \in \mathfrak{S}$, $\varphi|_A$ is $\sigma(H,G)$-continuous on the compact set A; it follows that

$$\sup_{h \in A} |\langle h,\varphi\rangle| < \infty$$

and $\tau_{\mathfrak{S}}$ is a locally convex topology on $\hat{G}$. Since a pointwise limit of linear functionals is linear, and the uniform (on A) limit of continuous functions is continuous, $\hat{G}$ is $\tau_{\mathfrak{S}}$-complete.

(d) $\mathfrak{S}$ is a cover of H by absolutely convex sets, which are $\sigma(H,G)$- and $\sigma(H,\hat{G})$-compact. The Mackey-theorem on compatible topologies gives

$$(G,\tau_{\mathfrak{S}})' = H = (\hat{G},\tau_{\mathfrak{S}})'$$

which, by the Hahn-Banach-separation-theorem, implies that G is $\sigma(\hat{G},H)$-dense = $\tau_{\mathfrak{S}}$-dense in $\hat{G}$. ∎

If E is a locally convex space, $\mathfrak{S} = \{U^o \subset E' \mid U \in \mathfrak{U}_E(o)\}$ then, by the theorem

$$\tilde{E} = \{\varphi \in E'^* \mid \text{for all } U \in \mathfrak{U}_E(o) : \varphi|_{U^o} \text{ is } \sigma(E',E)\text{-continuous}\}.$$

COROLLARY: A locally convex space E is complete if and only if every $\varphi \in E'^*$ whose restrictions to equicontinuous sets are $\sigma(E',E)$-continuous, is $\sigma(E',E)$-continuous (and therefore represented by an element in E).

This will be referred to as GROTHENDIECK'S COMPLETENESS CRITERION.

0.7. Some information on compact, convex sets will be useful: A subset D of a convex set K is called <u>extremal</u> if

$$x,y \in K,\ \alpha > o,\ \beta > o,\ \alpha+\beta = 1,\ \alpha x + \beta y \in D$$

implies $x,y \in D$. A point $x_o \in K$ is an <u>extreme point</u> if the set $\{x_o\}$ is extremal. $\operatorname{ext} K$ stands for the set of all extreme points of K.

If K is compact and convex in a locally convex space E, an argument with Zorn's lemma shows that every closed non-empty, extremal subset of K contains an extreme point. Since for every upper-semi-continuous, convex $f: K \to \mathbb{R}$ the set

$$\{x \in K \mid f(x) = \sup_{y \in K} f(y)\}$$

is closed, extremal, and non-empty, every such function attains its supremum in an extreme point (BAUER'S MAXIMUM-PRINCIPLE). In particular, this is true for every $\varphi \in E'$: the Hahn-Banach-separation-theorem implies that

$$\overline{\text{co ext } K} = K$$

(KREIN-MILMAN-THEOREM). If L is compact, then the extreme points of $\overline{\text{co } L}$ are in L (MILMAN-THEOREM).

0.8. Finally, it is necessary to mention a class of locally convex spaces which appear in many applications: If $E_1 \hookrightarrow E_2 \hookrightarrow \dots$ is a sequence of continuously embedded Fréchet-spaces, then

$$E := \bigcup_{n=1}^{\infty} E_n$$

is a vector space, which is called an <u>(LF)-space</u> once it is equipped with the finest locally convex topology which makes all the embeddings $E_n \hookrightarrow E$ continuous. If either the linking mappings are open (onto its image: <u>strict</u> (LF)-spaces) or <u>weakly compact</u> (there is neighbourhood in E_n which is $\sigma(E_{n+1}, E'_{n+1})$-relatively compact) E is separated, complete, and every bounded set is bounded in some step E_n.

§1. COUNTABLY COMPACT SETS AND THE THEOREM OF EBERLEIN - GROTHENDIECK

1.1. In dealing with compact subsets of non-metric spaces usually two difficulties arise. First, to test compactness it does not suffice to find limit-points of countable nets, and second, a sequence in a compact set need not have a convergent subsequence. All topological spaces are supposed to be separated.

DEFINITION: Let A be a subset of a topological (Hausdorff) space X

(1) A is countably compact if every sequence in A has a cluster-point in A .

(2) A is sequentially compact if every sequence in A has a convergent subsequence with limit in A .

(3) A is relatively countably compact if every sequence in A has a cluster-point in X .

(4) A is relatively sequentially compact if every sequence in A has a convergent subsequence (with limit in X) .

The definition of a relatively countably compact set A is not that the closure $\bar{A}$ is countably compact: it will be part of the investigation under which circumstances this is true (see example 1.2.(9) below).

1.2. Some facts on the relation of these notions: It is easy to see that

(1) Every (relatively) compact set is (relatively) countably compact.

(2) Every (relatively) sequentially compact set is (relatively) countably compact.

It will turn out that these are the only (non-trivial) implications which hold generally (see (6) and (8)).

(3) In a uniform space every relatively countably compact set is precompact. In particular, a relatively countably compact subset of a locally convex space is bounded.

Proof: Assume there is a uniform symmetric neighbourhood U and a sequence (x_n) in A such that

$$x_{n+1} \notin \bigcup_{i=1}^{n} U(x_i)$$

If x is a cluster point of (x_n), V a uniform symmetric neighbourhood with $V \circ V \subset U$ and $x_m \in V(x)$ then

$$V(x) \subset U(x_m)$$

which implies that no other x_n is contained in $V(x)$: contradiction. ■

In a uniform space, a set is compact if and only if it is complete and precompact. So (3) yields

(4) In a uniform space a subset is relatively compact if and only if it is relatively countably compact and its closure is complete.

A theorem of Hausdorff states that

(5) In metric spaces: compact = countably compact = sequentially compact; the same is true for the relative notions.

Now some counterexamples: the closed unit ball in $(\ell^\infty)'$ is $\sigma((\ell^\infty)', \ell^\infty)$-compact. But it is not sequentially compact: If the sequence (e_n) of unit vectors had a convergent subsequence $(e_n)_{n \in \mathbb{N}_1}$ this implied that

$$(\langle e_n, (\xi_m)\rangle = \xi_n)_{n \in \mathbb{N}_1}$$

would converge for all bounded sequences (ξ_m). Note that by the theorem on bipolars span$\{e_n\}$ in $(\ell^\infty)'$ is $\sigma((\ell^\infty)', \ell^\infty)$-dense:

(6) There is a separable locally convex space with a compact absolutely convex subset which is not sequentially compact.

But also

(7) There is a sequentially complete locally convex space with an absolutely convex, closed, sequentially compact (and so countably compact) subset which is not compact.

Take for this

$$E: = \{(\xi_r) \in \mathbb{R}^{\mathbb{R}} \mid \xi_r = 0 \text{ except for countably many } r \in \mathbb{R}\}$$

with the topology induced by $\mathbb{R}^{\mathbb{R}}$. The set

$$A: = \{(\xi_r) \in E \mid |\xi_r| \leq 1\}$$

is closed and sequentially compact, since a sequence (x_n) in A is already situated in some metrizable subspace $\mathbb{R}^D \subset E$, D countable. But it is not compact, because the closure of A in $\mathbb{R}^{\mathbb{R}}$ is

$$\tilde{A} = \{(\xi_r) \in \mathbb{R}^{\mathbb{R}} \mid |\xi_r| \leq 1\} \not\subset E .$$

This observation will be used frequently:

(8) If $A \subset X \subset Y$ and A is relatively compact in Y, then A is relatively compact in X if and only if

$$\bar{A}^{Y} \subset X$$

The foregoing example can be refined to yield

(9) There is a locally convex space with an absolutely convex, sequentially compact subset whose closure is not countably compact.

For a disjoint sequence of uncountable sets X_n put $X: = \bigcup_{n=1}^{\infty} X_n$, define for $f: X \to \mathbb{R}$ the support

$$\operatorname{supp} f: = \{x \in X \mid f(x) \neq 0\} ,$$

and take

$$E: = \{f: X \to \mathbb{R} \mid \exists\, n \in \mathbb{N}: \operatorname{supp} f \cap \bigcup_{m=n}^{\infty} X_m \text{ countable}\}$$

with the topology of pointwise convergence on X, i.e. the induced topology from $\mathbb{R}^X$. The set

$$A: = \{f \in E \mid \text{supp } f \text{ countable, } \|f\|_\infty \leq 1\}$$

is, as before, sequentially compact; in the closure

$$\bar{A}^E = \{f \in E \mid \|f\|_\infty \leq 1\} ,$$

however, the sequence consisting of the characteristic functions χ_{X_n} of X_n converges in $\mathbb{R}^X$ to $\chi_X \notin E$ and , for this, has no cluster-point in E . ∎

1.3. Note that all the counterexamples were actually locally convex spaces in a certain weak topology. The investigation of these different notions of compactness in $(E, \sigma(E, E'))$ will be split up in two parts:

(a) When is a $\sigma(E, E')$ — relatively countably compact subset $\sigma(E, E')$-relatively compact: theorem of Eberlein

(b) When do all the notions collapse in the finest way: angelic spaces, theorem of Eberlein - Šmulian (§3)

The sufficient condition, stated in 1.2.(4), that a countably compact set is compact if it is complete is close to being tautological in the weak topology of a locally convex space E since a $\sigma(E, E')$- complete bounded set is always $\sigma(E, E')$-compact; E is already semi-reflexive if all bounded convex closed sets are $\sigma(E, E')$-complete. On the other hand, completeness (and so compactness) of a convex set A in the weak topology $\sigma(E, E')$ implies that it is complete in all locally convex topologies which are compatible with the dual system $\langle E, E' \rangle$. The theorem of Eberlein - Grothendieck will state, that completeness in such a finer, compatible topology already suffices for question 1.3(a).

1.4. Part of this theorem can best be formulated in the framework of spaces of continuous functions.

A. Grothendieck introduced the following notion, which seems to be technical at first glance but will turn out as being an extremely useful tool to characterize relatively compact sets.

DEFINITION: Let Z be a topological (Hausdorff)-space, X a set, and $A \subset Z^X$ a set of functions $X \to Z$. Then X and A have the <u>interchangeable double-limit-property</u> (in Z), in signs: $A \sim X$ (in Z), if for every sequence (x_n) in X and every sequence (f_m) in A

$$\lim_m \lim_n f_m(x_n) = \lim_n \lim_m f_m(x_n)$$

whenever all the limits involved exist.

$A \sim X$ without "(in Z)" will be used if it is clear in which space Z the limits are; in the case of real functions some attention is to be paid to the cases $Z = \mathbb{R}$ and $Z = \bar{\mathbb{R}} := [-\infty, +\infty]$; both will occur. Since the property is symmetric in A and X, also $X \sim A$ will be written sometimes.
For topological (Hausdorff) spaces X and Z, define

$$C(X,Z) := \{f: X \to Z \mid \text{continuous}\}$$

and ω_X the (Hausdorff)-<u>topology of pointwise convergence</u> in X <u>on $C(X,Z)$</u>, i.e. the topology which is induced by Z^X.

If X is countably compact and $A \subset C(X,Z)$ is ω_X-relatively countably compact, then (x_n) in X and (f_m) in A have cluster-points $x \in X$ and $f \in C(X,Z)$ respectively: $f(x_n)$ is therefore a cluster-point of $(f_m(x_n))_m$, but if

$$\lim_m f_m(x_n)$$

exists, the only cluster-point of this sequence is the limit:

$$\lim_m f_m(x_n) = f(x_n) .$$

This argument repeated implies that

$$\lim_n \lim_m f_m(x_n) = f(x) = \lim_m \lim_n f_m(x_n)$$

if all these limits exist: $A \sim X$ (in Z) .

This is the first part of the following

THEOREM: Let D be a dense subset of a countably compact space X, Z a compact metric space and $A \subset C(X,Z)$. The following are equivalent:

(1) A is ω_X-relatively countably compact in $C(X,Z)$.

(2) $A \sim D$ (in Z).

(3) A is ω_X-relatively compact in $C(X,Z)$.

In particular $A \sim X$ is equivalent to each of the conditions (1) - (3).

Actually the implication

(1) $\Rightarrow$(2) holds for D relatively countably compact, Z arbitrary (this is clear by the foregoing remark; see Ex. 1.18.)

(2) $\Rightarrow$(3) holds for arbitrary X, Z compact metric (but see Ex. 1.17.(d))

(3) $\Rightarrow$(1) holds always.

(1) $\Rightarrow$ (3) holds for various other X and Z (see Ex. 1.21., 3.7. and 3.15.).

For the remaining implication (2) $\Rightarrow$(3) the following characterization of continuous functions will be helpful.

LEMMA: If D is a dense subset of a topological space X, and Z a metric space, then $f: X \to Z$ is continuous if and only if for all $x \in X$

$$\lim_{\substack{y \to x \\ y \in D}} f(y) = f(x)$$

Proof: Given an $x_0 \in X$ and an $\varepsilon > 0$ then there is an open neighbourhood U of x_0 with $f(U \cap D) \subset B(f(x_0),\varepsilon)$: = the ball with radius ε and center $f(x_0)$. If $x \in U$ there is an open neighbourhood V of x with $f(V \cap D) \subset B(f(x), \varepsilon)$; but $U \cap V \neq \emptyset$ and open, so there exists a $y \in U \cap V \cap D$ and the triangle inequality yields

$$d(f(x), f(x_0)) \leq d(f(x), f(y)) + d(f(y), f(x_0)) \leq 2\varepsilon \quad . \quad \blacksquare$$

(It is obvious how to change this proof for regular spaces Z)

Proof of the theorem's implication (2)$\leadsto$(3): The space Z^X is compact, so it suffices to show that every $f \in \bar{A}^{Z^X}$ is continuous (that is applying 1.2.(8)). Assume it is not, then, by the lemma, there is an $x_0 \in X$ and an open neighborhood U of $f(x_0)$, such that

(*) $$\forall_{V \in \mathfrak{U}(x_0)} \quad \exists_{x \in V \cap D} \quad f(x) \notin U$$

Thus for an arbitrary $f_1 \in A$ - which is continuous - there is an $x_1 \in D$ with

$$d(f_1(x_1), f_1(x_0)) \leq 1$$
$$f(x_1) \notin U$$

Since f is in the closure of A, there is an $f_2 \in A$ with

$$d(f_2(x_i), f(x_i)) \leq \tfrac{1}{2} \quad i = 0,1 \ ;$$

f_2 is continuous, so (*) ascertains the existence of an $x_2 \in D$ with

$$d(f_j(x_2), f_j(x_0)) \leq \tfrac{1}{2} \quad j = 1,2$$
$$f(x_2) \notin U$$

Proceeding by induction, there are $x_n \in D$, $f_n \in A$ with

(a) $d(f_n(x_i), f(x_i)) \leq \frac{1}{n}$ $\qquad i = 0,1,\dots,n-1$

(b) $d(f_j(x_n), f_j(x_0)) \leq \frac{1}{n}$ $\qquad j = 1,\dots,n$

$f(x_n) \notin U$

By the sequential compactness of Z the sequence $(f(x_m))_m$ may be assumed to be converging to some $z_0 \notin U$. Now (a) says

$$\lim_n f_n(x_i) = f(x_i)$$

so

$$\lim_m \lim_n f_n(x_m) = z_0$$

On the other hand (b) implies

$$\lim_m f_j(x_m) = f_j(x_0)$$

and (a) for $i = 0$

$$f(x_0) = \lim_n f_n(x_0) = \lim_n \lim_m f_n(x_m) \quad .$$

But $f(x_0) \neq z_0$, therefore a contradiction to the interchangeable double limit property is established. ■

Under the assumption that Z is compact, $A \subset C(X,Z)$ being ω_X - relatively compact means exactly

(3') the pointwise limit of functions in A is continuous.

(again this is the argument 1.2.(8)). The theorem will be improved considerably in §3.

1.5. Specializing to real-valued functions $(C(X) := C(X,\mathbb{R}))$

COROLLARY: Let X be a topological space, $D \subset X$ dense.

(1) If $A \subset C^b(X)$, the space of bounded continuous functions, is uniformly bounded and $A \sim D$ (in $\mathbb{R}$), then A is ω_X - relatively compact in $C^b(X)$.

(2) If $A \subset C(X)$ is pointwise bounded and $A \sim D$ (in $\overline{\mathbb{R}}$), then A is ω_X - relatively compact in $C(X)$.

Proof: (1) If $|f(x)| \leq C$ for all $x \in X$ and $f \in A$ then, by the theorem, remark on $(2) \curvearrowright (3)$, A is ω_X - relatively compact in

$$C(X, [-C, +C])$$

(2) Consider $C(X)$ as a subset (in its topology) of $C(X,\overline{\mathbb{R}})$ and notice that pointwise limits of functions in A are everywhere finite: this means that the ω_X-closure of A in $C(X,\overline{\mathbb{R}})$, which is compact by the theorem, is already in $C(X)$. ■

(1) is false when "uniformly bounded" is replaced by "pointwise bounded" (see Ex. 1.16.).

1.6. Considering elements x of a locally convex space E as functions on equicontinuous subsets of E' and applying Grothendieck's completeness criterion will yield the

THEOREM (W.F.Eberlein-A.Grothendieck): Let (E,τ) be a locally convex space, $B \subset E$ convex and τ-complete. For a subset $A \subset B$ the following are equivalent:

(1) A is $\sigma(E,E')$-relatively countably compact.

(2) A is $\sigma(E,E')$-relatively compact.

(3) A is bounded and has the interchangeable double-limit-property (in $\mathbb{R}$) with all τ-equicontinuous subsets of E'.

(1)$\Rightarrow$(3) holds without the completeness assumption. (1)$\Leftrightarrow$(2) is usually called Eberlein-theorem; it will be seen (3.10.) that sometimes, e.g. in metrizable spaces, the completeness assumption can be deleted - but in general it is false for sequentially complete spaces (1.2.(7)). (1)$\Leftrightarrow$(2) does not hold without "relative": take for an example the set of 1.2.(7) in the completion of the space. Furthermore, note that the finer the compatible topology τ is in which B is complete the more has to be checked in (3) - in case that τ is the Mackey-topology $\mu(E,E')$, that is the weakest assumption of completeness, $A \sim Q$ for all convex $\sigma(E',E)$-compact sets; in case that τ is already $\sigma(E,E')$, the double limit-property is redundant. The equivalences hold in particular for quasicomplete spaces.

Proof: The weak closure of A being in B, without loss of generality E can be assumed to be complete.

In all cases A is bounded, so there is for every zero-neighbourhood U a $\lambda > 0$ such that (U° carries $\sigma(E',E)$ and is therefore compact)

$$\kappa_{U^{\circ}} : (\bar{A},\sigma(E,E')) \longrightarrow (C(U^{\circ},[-\lambda,\lambda]),\omega_{U^{\circ}})$$
$$x \rightsquigarrow x|_{U^{\circ}}$$

is well-defined and continuous.

(1)$\curvearrowright$(3): $\kappa_{U^o}(A)$ is ω_{U^o} - relatively countably compact and therefore, by the theorem 1.4 above,

$$\kappa_{U^o}(A) \sim U^o$$

(3)$\curvearrowright$(2): First observe that, again by this theorem, $\kappa_{U^o}(A)$ is ω_{U^o} - relatively compact and so, by 1.4.(3'), the pointwise (on U^o) limit of elements $x|_{U^o}$, $x \in A$, is $\sigma(E',E)$ continuous. This yields that for

$$z \in \bar{A}^{\sigma(E'^*,E')}$$

all the restrictions $z|_{U^o}$ are $\sigma(E',E)$-continuous: Grothendieck's completeness theorem (0.6.) says $z \in E$. ■

It is clear that property (3) has only to be checked for equicontinuous sets U^o, $U \in \hat{\mathfrak{U}}$, if

$$\{ \varepsilon U \mid \varepsilon > o \ , \ U \in \hat{\mathfrak{U}} \}$$

is a basis of neighbourhoods - for example, only for the dual unit ball of a Banach-space. Furthermore, according to theorem 1.4., it is only necessary to check the double-limit-property for a dense subset of U^o: so assume $F \subset E'$ is a subspace such that

$$\overline{F \cap U^o}^{\sigma(E',E)} = U^o$$

for all $U \in \hat{\mathfrak{U}}$. Then the following is also an equivalent condition:

(4) A is bounded and $A \sim Q$ for all equicontinuous subsets Q of F .

An important example for this is constructed with the aid of the Krein-Milman-theorem: the closed convex hull of the extreme points ext U^o of U^o is dense in U^o, therefore

$$F := \text{span} \bigcup_{U \in \hat{\mathfrak{U}}} \text{ext } U^o$$

is such a subspace $F \subset E'$. The topology $\sigma(E,F)$ will be investigated later (§8) . Furthermore, again by theorem 1.4., it is enough that the underlying space X of the continuous functions is countably compact:

(5) A is bounded and $A \sim Q$ for all convex $\sigma(E',E)$-relatively countably compact subsets Q of E' .

again is equivalent, once it is established that a mapping κ_Q is well-defined, that is applying the following lemma to $G = (E', \sigma(E',E))$ and the barrel A^o.

LEMMA: Every convex, relatively countably compact subset Q of a locally convex space G is contained in a Banach-disc $D \subset G$. In particular, Q is absorbed by every barrel in G.

Proof: Since Q is bounded, the map ($\tilde{G}$ the completion)

$$T: \ell^1(Q) \longrightarrow \tilde{G}$$
$$(\xi_q) \rightsquigarrow \sum_q \xi_q \, q$$

is well-defined, continuous and therefore $T(U)$, U the unit ball in $\ell^1(Q)$, is a Banach-disc in $\tilde{G}$ containing Q. But actually, $T(U) \subset G$: For this take $\xi_n > 0$, $\sum_{n=1}^{N} \xi_n =: s_N \rightarrow s$, $x_n \in Q$, then

$$\sum_{n=1}^{\infty} \xi_n x_n = \lim_N s_N \sum_{n=1}^{N} \frac{\xi_n}{s_N} x_n = s \lim_N \underbrace{\sum_{n=1}^{N} \frac{\xi_n}{s_N} x_n}_{\in Q} \in G$$

since Q is convex and relatively countably compact. But now a split in positive and negative parts yields $T(U) \subset G$. ∎

1.7. The Eberlein - Grothendieck theorem gives new characterizations of reflexivity:

COROLLARY 1: For a quasicomplete locally convex space E are equivalent:

(1) E is semi-reflexive.

(2) all bounded subsets are $\sigma(E,E')$ - relatively countably compact.

(3) all closed separable subspaces are semi-reflexive.

1.8. The theorem also provides some information on countable compactness for other compatible topologies:

COROLLARY 2: Let B be a convex $\mu(E,E')$-complete subset of a locally convex space (E,τ), then $A \subset B$ is τ-relatively countably compact if and only if it is τ-relatively compact.

Proof: If A is τ-relatively countably compact, then it is τ-precompact (1.2.(3)) and $\sigma(E,E')$-relatively countably compact = $\sigma(E,E')$-relatively compact by the theorem. But this implies that $\bar{A}^{\sigma(E,E')}$ is $\sigma(E,E')$-complete in particular

τ-complete: $\overline{A}^{\tau}$ is τ-precompact and τ-complete and therefore τ-compact. ■

1.9. Surely, the assumption that A is in some convex complete set B can be always replaced by the condition that the closed convex hull $\overline{\text{co}\, A}$ is $\mu(E,E')$-complete or complete in any compatible topology. It is not true that $\overline{\text{co}\, A}$ is compact if A is: Krein's theorem will state that this is exactly the case if $\overline{\text{co}\, A}$ is $\mu(E,E')$-complete (see § 7 and Ex. 1.14.)

EXERCISES

1.10 A set is not countably compact if and only if it has a countable discrete closed subset.

1.11 Give a characterization of (relatively) countable compact sets in terms of countable subsets.

1.12 If a convex subset of a locally convex space is compact (sequentially compact, relatively ...) so is its absolutely convex hull.

1.13 Show that the absolutely convex hull of a precompact subset of a locally convex space is precompact.

1.14. Prove:the closed convex hull of a compact convex set is compact if and only if it is complete.

1.15. Let $\langle G,H\rangle$ be a dual system of real vector spaces and $K \subset G$ a $\sigma(G,H)$-bounded and convex set. Then K is $\sigma(G,H)$- compact if and only if for each $\varphi \in H^*$ with

$$\langle \varphi,h\rangle \leq \sup_{x \in K} \langle h,x\rangle \quad \text{for all} \quad h \in H$$

there is an $x_o \in K$ with

$$\langle \varphi,h\rangle = \langle h,x_o\rangle \quad \text{for all} \quad h \in H \quad .$$

This is a result of V.L. Šmulian. (Hint: Consider $K \subset H^*$, Hahn - Banach - separation-theorem and 1.2.(8)).

1.16. (a) For

$$f_n: \begin{array}{ccc} \mathbb{N} & \longrightarrow & \mathbb{N} \\ \Downarrow & & \Downarrow \\ m & \rightsquigarrow & \min(m,n) \end{array}$$

show that $A: = \{f_m\} \subset C^{b}(\mathbb{N}, \mathbb{R})$ has $A \sim \mathbb{N}$ (in $\mathbb{R}$).
(Hint: Embed into $C(\mathbb{N} \cup \{\infty\}, \overline{\mathbb{R}})$.)

(b) Conclude from (a) that a pointwise bounded subset $A \subset C^{b}(X)$ with $A \sim X$ (in $\mathbb{R}$) need not be ω_X-relatively compact in $C^{b}(X)$.

1.17. (a) Show that $\mathbb{R} \sim \mathbb{R}' = \mathbb{R}$ (in $\mathbb{R}$).

(b) Give an example of a subset A of a complete locally convex space which is unbounded but has interchangeable double-limits (in $\mathbb{R}$) with all equicontinuous sets.

(c) On the other hand, if $A \sim U^o$ (in $\overline{\mathbb{R}}$), for all equicontinuous sets U^o, then A is bounded.

(d) The implication (2)$\rightsquigarrow$(3) in theorem 1.4. does not hold provided X is compact, Z metric and only locally compact (Hint: $\mathbb{R} \subset C([0,1], \mathbb{R})$.)

1.18. Show that with

$$f_c(x) := \frac{|cx|}{1+|cx|} \quad x, c \in \mathbb{R}$$

the set

$$A := \{f_c \mid 0 \le c \le 1\} \subset C(\mathbb{R}, [0,1])$$

is $\omega_{\mathbb{R}}$-compact but $A \sim \mathbb{R}$ does not hold.

1.19. In the statements of 1.4 and 1.5 the topology of X can be replaced by the coarsest topology τ_A such that all functions $f \in A$ are continuous (τ_A might not be Hausdorff). For example, in 1.4. an additional equivalence is

$$\overline{A}^{Z^X} \subset C((X,\tau_A), Z) \quad [\subset C(X,Z)] \quad .$$

1.20. Show that for the implication (3)$\rightsquigarrow$(1) in the Eberlein - Grothendieck - theorem the completeness assumption cannot be just omitted. (Hint: Take the unit ball of a normed space with reflexive completion.)

1.21. (a) If X is countably compact, Z metric, then a subset $A \subset C(X,Z)$ is ω_X-relatively countably compact if and only if it is ω_X-relatively compact. (Hint: Start as in the proof of (2)$\rightsquigarrow$(3) of theorem 1.4., then take a limit-point $g \in C(X,Z)$ of the sequence (f_n), observe that

$g(x_m) = f(x_m)$ and use a limit point of (x_m) to show that $(f(x_m))_m$ may be assumed to converge.)

(b) If Z is metric, X a topological space such that a function $f: X \to Z$ is continuous if and only if

(1) it is sequentially continuous, or

(2) all restrictions to (countably) compact sets are continuous.

then every ω_X-relatively countably compact subset of $C(X,Z)$ is ω_X-relatively compact. (Hint: Take f in the Z^X-closure of A and show that $f|_K$, K compact or a converging sequence, is continuous.)

The conclusion holds in particular, if X is metric or locally compact

1.22. One of the original approaches of A. Grothendieck to the Eberlein-theorem runs as follows:

(a) Let H be a vector-space, $E \subset H$ a subspace, and $C \subset E$ a convex subset. If μ and τ are locally convex (separated) topologies on H with $(H,\mu)' = (H,\tau)'$, then C is τ-relatively compact in E if and only if $\overline{C}^{\tau} \cap E$ is μ-complete and C is τ-relatively compact in H.

(b) If $T_i : E \to E_i$ are linear mappings, μ_i and τ_i locally convex (separated) topologies on E_i with $(E_i,\tau_i)' = (E_i,\mu_i)'$, and if τ and μ, the initial topologies of (τ_i) and (μ_i) respectively, are separated, then a convex $C \subset E$ is τ-relatively compact if and only if $\overline{C}$ is μ-complete and all $T_i(C)$ are τ_i-relatively compact.

(c) Show with (b) that the proof of Eberlein's theorem can be reduced to the case of Banach-spaces.

1.23 Give examples when $\overline{C(X,Z)}^{Z^X} = Z^X$.

1.24. For a vector-space $E \subset \mathbb{R}^X$ and a subset $D \subset X$, the topology ω_D of pointwise convergence on E is the weak-topology $\sigma(E, \operatorname{span} D)$, $\operatorname{span} D \subset E^*$ via Dirac-functionals.

§2. BOUNDING SETS IN THE WEAK TOPOLOGY

2.1. Call a subset A of a topological space $(X,\mathcal{T})$ $\mathcal{T}$-bounding, if all $\mathcal{T}$-continuous functions $X \to \mathbb{R}$ are bounded on A. For $A = X$, topologists speak of pseudocompact spaces. Pseudocompact subsets (in their induced topology) are bounding. It is easily seen, that relatively countably compact sets are bounding, and that bounding subsets of uniform spaces are precompact (Ex. 2.9.): but there are pseudocompact spaces which are not compact: The ordinal space $[1,\omega_1[$ consisting of all ordinal numbers strictly smaller than the first uncountable one ω_1, equipped with the order topology, is countably compact and therefore pseudocompact, but not compact; Tychonoff's Plank

$$P := [1,\omega_1] \times [1,\omega_o] \setminus (\omega_1,\omega_o)$$

(ω_o the ordinal number of $\mathbb{N}$) is not even countably compact (since $(\omega_1,n) \to (\omega_1,\omega_o)$) but pseudocompact: If a continuous function $f : P \to \mathbb{R}$ admits points (x_n,k_n) with $f((x_n,k_n)) \geq n$, then necessarily $(x_n,k_n) \to (\omega_1,\omega_o)$ which by the very nature of ω_1 is only possible if $x_n = \omega_1$ eventually. The continuity ascertains the existence of $y_n < \omega_1$ with

$$f\Big|_{[y_n,\omega_1]\times\{k_n\}} \geq n - 1 \quad ;$$

but this yields for a $y < \omega_1$ with $y_n \leq y$ for all n,

$$f(y,\omega_o) = \lim_{n\to\infty} f(y,k_n) = \infty$$

— a contradiction.

To transfer these counterexamples to locally convex spaces, first note the simple

LEMMA: If a completely regular Hausdorff space X is considered as a subset of $(C(X))^*$ and $D \subset X$, then $D = X \cap \operatorname{span} D$. Furthermore, the embedding

$$X \hookrightarrow (C(X))^*$$

is topological, if $(C(X))^*$ is equipped with $\sigma((C(X))^*, C(X))$.

If X is compact, this implies that D is closed in span D with the topology σ(span D, C(X)) . In particular, the pseudocompact, not countably compact Tychonoff-Plank P in $X := [1, \omega_1] \times [1, \omega_0]$ is closed and not countably compact in span P with σ(span P, C(X)) :

PROPOSITION: There are locally convex spaces (in their weak topology) with pseudocompact (in particular: bounding) closed subsets which are not countably compact.

For another example see Ex. 2.15. .

2.2. A theorem due to L. Nachbin and T. Shirota (see J.Schmets [46]) states, that in a completely regular Hausdorff-space X every bounding set is relatively compact if and only if C(X) with the topology of uniform convergence on all compact sets, the compact-open-topology, is barrelled. (For the easy part of the proof of this theorem see Ex. 2.14.)

2.3. Under which circumstances is a bounding set in a locally convex space relatively compact? The general case will be reduced to the weak topology (see Corollary 4 in 2.7.).

THEOREM (M. Valdivia): If E is a locally convex space with $\sigma(E',E)$-separable dual, then every $\sigma(E,E')$-bounding subset of E is $\sigma(E,E')$-relatively compact.

The proof is based on the following

LEMMA: If E' is $\sigma(E',E)$-separable, $z \in E'^* \setminus E$ and $F := E \oplus [z]$, then E' is also $\sigma(E',F)$-separable.

Proof of the lemma (W. Govaerts):

(a) There is a countable $\sigma(E',E)$-dense subset $D \subset E'$ with

$$B(D) := \bigcap_{\varphi \in D} \{y \in E \mid \langle \varphi,y \rangle = \langle \varphi,z \rangle\} = \emptyset \quad :$$

For this, take the countable $\sigma(E',E)$-dense subset $D_1 \subset E'$ which exists by assumption and assume $B(D_1) \neq \emptyset$: However, it cannot contain two different points y_1, y_2 since there were a $\varphi \in D_1$ with $\langle \varphi,y_1 \rangle \neq \langle \varphi,y_2 \rangle$. If $B(D_1) = \{y_0\}$, then there is a $\varphi_0 \in E'$ with $\langle \varphi_0,z \rangle \neq \langle \varphi_0,y_0 \rangle$ because $z \neq y_0$ in E'^*. But then $B(D_1 \cup \{\varphi_0\}) = \emptyset$.

(b) To show that span D is $\sigma(E',F)$-dense, take an $x + \lambda z \in F = (E', \sigma(E',F))'$ and assume

$$\langle \varphi, x + \lambda z \rangle = 0 \quad \text{for all} \quad \varphi \in D .$$

If $\lambda \neq 0$, then this implies

$$\frac{x}{-\lambda} \in B(D)$$

which is impossible, so $\lambda = 0$ and

$$\langle \varphi,x \rangle = 0 \quad \text{for all} \quad \varphi \in D :$$

But this means $x = 0$. ∎

It is immediate that the proof carries over to densities of any other (infinite) cardinality.

Proof of the theorem: A bounding subset is, by Mackey's theorem, bounded. So, by a now familiar argument, it is to prove that every

$$z \in \bar{A}^{\sigma(E'^*,E')}$$

is in E . Assume not and define $F := E \oplus [z]$; then the lemma ascertains the existence of a $\sigma(E',F)$-dense sequence (φ_n) in E' , i.e. for $y \in F$

$$\langle \varphi_n, y - z \rangle = 0 \quad \text{for all} \quad n = 1,2,\ldots$$

if and only if $y = z$. The function

$$f(y) := \sum_{n=1}^{\infty} \frac{1}{2^n} \frac{|\langle \varphi_n, y - z \rangle|}{1 + |\langle \varphi_n, y - z \rangle|} \qquad y \in F$$

is obviously $\sigma(F,E')$-continuous, nowhere zero on E , and

$$\inf_{x \in A} f(x) = 0$$

— the latter since $z \in \bar{A}^{\sigma(F,E')}$. But this shows that $\frac{1}{f}$ is $\sigma(E,E')$-continuous on E and unbounded on A : Contradiction. ■

In particular, this proposition holds for separable normed spaces.

2.4. The non-separable case, by the examples in 2.1., requires an additional assumption.

COROLLARY 1 (M. Valdivia): In a quasicomplete locally convex space E a subset is $\sigma(E,E')$-relatively compact if and only if it is $\sigma(E,E')$-bounding.

Proof: Applying the double-limit-criterion for weak compactness, take an equicontinuous sequence (φ_n) in E' and a sequence (x_m) in A with existing double limits. If

$$H := \overline{\operatorname{span} \{\varphi_n\}}^{\sigma(E',E)}$$

then the weak topology of the quotient

$$\kappa : (E,\tau) \to (E,\tau)/H^o =: (G,\kappa(\tau))$$

is $\sigma(G,H)$, the dual H is $\sigma(H,G)$-separable, $\kappa(A)$ is $\sigma(G,H)$-bounding and therefore by the theorem, $\kappa(A)$ is $\sigma(G,H)$ - relatively compact. The sequence (φ_n) is obviously equicontinuous in $(G,\kappa(\tau))' = H$ such that the Eberlein-Grothendieck theorem (quasi-completeness was not necessary for this implication) gives

$$\lim_m \lim_n \langle\varphi_n, x_m\rangle_{E',E} = \lim_m \lim_n \langle\varphi_n, \kappa(x_m)\rangle_{H,G} = \lim_n \lim_m \ldots \quad . \quad ∎$$

2.5. Surely this implies

COROLLARY 2 (V. Pták): $\sigma(E,E')$-pseudocompact subsets of a quasicomplete locally convex space E are $\sigma(E,E')$-relatively compact.

2.6. The following reflexivity-criterion is an easy consequence of Corollary 1.

COROLLARY 3: A quasicomplete locally convex space E is semi-reflexive if and only if every $\sigma(E,E')$-continuous function on E is bounded on bounded sets.

The theorem tells that the assumption of quasicompleteness can be replaced by E' being $\sigma(E',E)$-separable. In particular, a separable normed space E is reflexive if all $\sigma(E,E')$-continuous functions are bounded on the unit ball.

2.7. The simple observation that in a uniform space every bounding subset is precompact (Ex. 2.9.) allows to treat others than the weak topology of a locally convex space:

COROLLARY 4: If A is a τ-bounding subset of a locally convex space (E,τ) and $\overline{\text{co}A}$ is $\mu(E,E')$-complete (or $(E',\sigma(E',E))$ is separable), then A is τ-relatively compact.

Proof: A is also $\sigma(E,E')$-bounding and so, by Corollary 1 (or the theorem), $\sigma(E,E')$-relatively compact : $\bar{A}^{\sigma(E,E')}$ is $\sigma(E,E')$-complete, in particular τ-complete; but A is also precompact. ∎

2.8. The Nachbin-Shirota-theorem together with Corollary 4 yields that for any locally convex space E which is quasicomplete in its Mackey-topology or has $\sigma(E',E)$-separable dual, the compact-open topology of $C(E)$ is barrelled.

EXERCISES

2.9. A bounding subset of a uniform space is precompact. (Hint: Take a uniform neighborhood V and a sequence (x_n) with $(V \circ V)(x_n) \cap (V \circ V)(x_m) = \emptyset$; $f_n(x_n) = 1$ and $f_n|_{\complement V(x_n)} = 0$.)

2.10. A subset A of a completely regular space X is bounding if and only if A considered as a subset of $C(X)^*$ is $\sigma(C(X)^*, C(X))$- bounded.

2.11. The unit ball of a normed space E is bounding if and only if $\dim E < \infty$.

2.12. A bounding subset of a metrizable space is relatively compact.

2.13. A topological space is pseudocompact if and only if every real-valued continuous function on it attains its supremum.

2.14. Let X be a completely regular space and $C_{co}(X)$ the space of continuous functions on X with the compact-open topology.

(a) If $A \subset X$ is bounding, then

$$\{f \in C(X) \mid \sup_{x \in A} |f(x)| \leq 1\}$$

is a barrel in $C_{co}(X)$.

(b) If $C_{co}(X)$ is barrelled, every bounding set in X is relatively compact.

2.15. Show that

$$A := \{f : \mathbb{R} \to \mathbb{R} \mid \text{supp } f \text{ countable}, \ f|_{\text{supp } f} = 1\}$$

is pseudocompact, sequentially compact, but not relatively compact in $E := \{f : \mathbb{R} \to \mathbb{R} \mid \text{supp } f \text{ countable}\}$ with the topology of $\mathbb{R}^{\mathbb{R}}$.

2.16. A pseudocompact normal topological space is countably compact (Hint: Use 1.10. and an extension theorem). Note that this, together with Proposition 2.1., implies that locally convex spaces are in general not normal topological spaces.

2.17. A pseudocompact subset of a quasicomplete locally convex space is relatively compact.

§3. SEQUENTIAL COMPACTNESS AND ANGELIC SPACES

3.1. If $f: X \to M$ is a continuous one-to-one mapping from a topological space into a metric one and if $A \subset X$ is relatively compact, then $f|_{\overline{A}}$ is a homeomorphism: in particular $\overline{A}$ is metrizable, A is relatively sequentially compact, every $x \in \overline{A}$ is the limit of a sequence in A, ... This simple principle will be generalized to relatively countably compact sets, and in this form it will constitute the basic idea of the joint investigation of all the compactness notions as well as of the question whether the sequential closure of a relatively countably compact set is its closure. Again, all topological spaces shall be separated.

LEMMA: Let X and Y be topological (Hausdorff) spaces, X regular (i.e. the closed neighbourhoods form a basis of all neighbourhoods of a point), and $\Phi: X \to Y$ continuous and injective. If $A \subset X$ is relatively countably compact and for all $B \subset \Phi(A)$ the sequential closure of B is closed, that is

$$\overline{B} = \{y \in Y \mid \exists (y_n) \text{ in } B \text{ with } y_n \to y\}$$

then $\Phi(\overline{A})$ is closed in Y and

$$\Phi|_{\overline{A}}$$

is a homeomorphism.

Sometimes this will be called the "angelic-lemma".

Proof: Assume $X = \overline{A}$ and $\overline{\Phi(A)} = Y$.

(a) First observe that a sequence in a topological space converges to x if and only if every subsequence has x as a cluster-point.

(b) If (x_n) is a sequence in A, $y \in \overline{\Phi(A)}$, and $\Phi(x_n) \to y$, then $y \in \Phi(\overline{A})$ and $x_n \to \Phi^{-1}(y)$: For this, take, according to (a), a subsequence $(x_n)_{n \in \mathbb{N}_1}$ of (x_n): A is relatively countably compact, so there is a cluster-point

x of $(x_n)_{n \in \mathbb{N}_1}$ and $\Phi(x)$ is also a cluster point of $(\Phi(x_n))_{n \in \mathbb{N}_1}$. But $\Phi(x_n)$ converges to y, therefore $\Phi(x) = y$.

(c) If $D \subset A$, then

$$\Phi(\bar{D}) \subset \overline{\Phi(D)} = \{\Phi(x) \mid \exists (x_n) \text{ in } D,\ \Phi(x_n) \to \Phi(x)\} \subset \Phi(\bar{D})$$

where the last inclusion holds by the result in (b): $\Phi(\bar{D}) = \overline{\Phi(D)}$. In particular $\Phi(\bar{A})$ is closed.

(d) Now take an arbitrary closed subset $C \subset \bar{A}$. For the proof of the lemma it suffices to show that $\Phi(C)$ is closed X is regular, so

$$C = \cap\{\bar{U} \mid C \subset U \text{ open}\}.$$

A being dense in $\bar{A} = X$ implies

$$\overline{U \cap A} = \bar{U}$$

for open sets U. By (c) it follows that

$$\Phi(C) = \cap\ \Phi(\bar{U}) = \cap\ \Phi(\overline{U \cap A}) = \cap \overline{\Phi(U \cap A)}$$

is closed. ∎

3.2. This yields quickly the

PROPOSITION: If E is a locally convex space with $\sigma(E',E)$-separable dual, then for every $\sigma(E,E')$-relatively countably compact subset $A \subset E$ the $\sigma(E,E')$-closure $\bar{A}^{\sigma}$ is metrizable.

Proof: For a countable dense subset $D \subset E'$ look at the map

$$\begin{aligned} \Phi: (E,\sigma(E,E')) &\longrightarrow (\mathbb{R}^D, \omega_D) \\ x &\rightsquigarrow (\langle \varphi, x\rangle)_{\varphi \in D} \end{aligned}$$

and apply the lemma. ∎

Therefore the compactness behaviour of metric-spaces carries over:

THEOREM (V.L.Šmulian): Let E be a locally convex space with $\sigma(E',E)$-separable dual. Then in $(E,\sigma(E,E'))$ the following statements hold:

(1) compact = countably compact = sequentially compact.

(2) relatively compact = relatively countably compact = relatively sequentially compact.

(3) If A is relatively (countably) compact, $x \in \bar{A}$, then there is a sequence (x_n) in A with $x_n \to x$.

(4) Every relatively countably compact, sequentially closed set is compact.

(5) If (x_n) is a sequence with cluster-point x and the set $\{x_n\}$ is relatively (countably) compact, then there is a subsequence converging to x.

According to the examples of §1, none of these statements, being more or less obvious in metric spaces, are generally true for weak topologies of locally convex spaces.

3.3. The Šmulian theorem holds in particular in separable normed spaces, in the $\sigma(E',E)$-dual of separable spaces - in particular in the (reflexive) space $\mathcal{D}'(\Omega)$ of distributions on an open set $\Omega \subset \mathbb{R}^n$ with the weak topology. To generalize it to a wider class of locally convex spaces, in particular to the weak topology of every metrizable space, again it is convenient to study first spaces of continuous functions. The fundamental notion, however, can be formulated even in arbitrary topological spaces. It is due to D.H. Fremlin and allows to exploit extensively the angelic-lemma.

DEFINITION: A topological (Hausdorff) space X is called angelic (or has countably determined compactness) if for every relatively countably compact set $A \subset X$ the following holds:

(a) A is relatively compact

(b) For each $x \in \bar{A}$ there is a sequence in A which converges to x .

Angelic spaces may deserve their name - or at least serve their purpose - as the following results show:

THEOREM:

(1) In angelic spaces all the statements (1)-(5) of the Šmulian theorem hold.

(2) If $\Phi: X \to Y$ is continuous and injective, X regular, and Y is angelic, then X is angelic.

Proof: (2) is immediate by the angelic-lemma.

For the first result, it is easily checked that in angelic spaces the statement (5) of Šmulian's theorem implies all the others. Therefore, take a sequence (x_n) with cluster-point x and assume that $\{x_n\}$ is relatively compact: If $x_n = x$ infinitely often, this subsequence converges to x . Hence suppose that $x_n \neq x$ for all n . In this case $x \in \overline{\{x_n\}}$ and, by the definition of an angelic space, there is a sequence (y_k) in $\{x_n\}$ which converges to x . However

$$y_k = x_{n_k}$$

such that a subsequence $(y_k)_{k \in \mathbb{N}_1}$ with $n_k < n_\ell$ for $k, \ell \in \mathbb{N}_1$, $k < \ell$ produces a subsequence of (x_n) which converges to x . ∎

3.4. In order to show under which circumstances function-spaces $C(X,Z)$ are angelic in the topology of pointwise convergence in the sequel some auxiliary statements are presented which, however, are also interesting for their own.

If $g: X \to \hat{X}$ is continuous with dense image, $h: \hat{Z} \to Z$ continuous and injective then the injective map

$$C(\hat{X},\hat{Z}) \longrightarrow C(X,Z)$$

$$f \rightsquigarrow h \circ f \circ g$$

is $\omega_{\hat{X}}$ - ω_X - continuous and therefore, if $C(X,Z)$ is ω_X-angelic then $C(\hat{X},\hat{Z})$, being certainly regular (if $\hat{Z}$ is), is also $\omega_{\hat{X}}$-angelic. This argument will be used frequently.

3.5. If $C(X,Z)$ is angelic then Z is (Ex. 3.14.) but surely this is not sufficient (Ex. 3.15.).

THEOREM (D.H.Fremlin): If $C(X,\mathbb{R})$ is ω_X-angelic, then $C(X,Z)$ is ω_X-angelic for every metric space Z.

Proof: If $A \subset C(X,Z)$ is relatively countably compact with respect to ω_X, that is the topology induced by Z^X, then the projection

$$A(x_o) := \{f(x_o) \mid f \in A\} \subset Z$$

is relatively (countably) compact in the metric space Z for all $x_o \in X$. The Tychonoff-theorem implies that $A \subset Z^X$ is relatively compact and it is to show that every

$$f \in \bar{A}^{Z^X}$$

is (a): continuous and (b): the ω_X-limit of a sequence in A.

(a) Fix $x_o \in X$, then

$$T: Z \longrightarrow \mathbb{R}$$
$$z \rightsquigarrow d(z, f(x_o))$$

is continuous, and so is

$$T_*: Z^X \longrightarrow \mathbb{R}^X$$
$$g \rightsquigarrow T \circ g$$

(with respect to the topologies ω_X). Since $T_*(C(X,Z)) \subset C(X,\mathbb{R})$, the set $T_*(A)$ is ω_X-relatively countably compact in the ω_X-angelic space $C(X,\mathbb{R})$, and therefore

$$d(f(\cdot),f(x_o)) = T_*f \in \overline{T_*(A)}^{\mathbb{R}^X} \subset C(X,\mathbb{R})$$

is continuous: But this means, that f is continuous in x_o.
(b) The idea to find the sequence is just the same. The map

$$S: C(X,Z) \longrightarrow C(X,\mathbb{R})$$
$$g \rightsquigarrow d(g(\cdot),f(\cdot))$$

is ω_X-continuous. $S(A)$ is therefore relatively compact in $C(X,\mathbb{R})$ and $0 = Sf \in S(\overline{A}) \subset \overline{S(A)}$. The angelic property ascertains the existence of a sequence (g_n) in A such that $(Sg_n(x))$ converges to $Sf(x) = 0$ for all $x \in X$, i.e.

$$g_n \to f \quad \text{in} \quad (C(X,Z),\, \omega_X) \quad . \qquad \blacksquare$$

3.6. The theorem 1.4. provides already a wide class of spaces $C(X,Z)$ where ω_X-relatively countably compact sets are ω_X-relatively compact. So it is crucial to find a class of spaces where the closure is the sequential closure.

THEOREM (M.De Wilde): Let X be a topological space, $X = \bigcup_{m=1}^{\infty} \overline{D}_m$, Z a compact metric space and $A \subset C(X,Z)$. If A has interchangeable double limits with all D_m: $A \sim D_m$ (in Z), then every

$$f \in \overline{A}^{Z^X}$$

is the ω_X-limit of a sequence in A.

(Part of this theorem can be found in the book of J. L. Kelley and I. Namioka.)
By theorem 1.4. the assumptions are fulfilled if A and all D_m are relatively countably compact in their respective topologies: Exactly this will be used later on.

Furthermore, note that again by theorem 1.4.(3') all the restrictions of f to $\overline{D}_m$ are continuous but f need not be continuous on the whole space: For this, take $Z := [0,1]$ and $X := \{\frac{1}{m} | m \in \mathbb{N}\} \cup \{0\}$ (the topologies induced by $\mathbb{R}$) and observe that in this case

$$\overline{C(X,Z)}^{Z^X} = Z^X \ .$$

Proof:

(a) Given functions $g_1,\dots,g_n \in C(\overline{D}_m,Z)$ and an $\varepsilon > 0$, then there is a finite set $L \subset D_m$ such that

$$\min_{y\in L}\max_{k\le n} d(g_k(x),g_k(y)) \le \varepsilon$$

for each $x \in \overline{D_m}$: This is true, since

$$\{G(x): = (g_1(x),\dots,g_n(x)) \in Z^n \,|\, x \in \overline{D_m}\}$$

is relatively compact in Z^n (with the maximum-metric) and therefore the open cover

$$G(\overline{D_m}) \subset \overline{G(D_m)} \subset \bigcup_{y\in D_m} B(G(y),\varepsilon)$$

admits a finite subcover.

(b) The result of (a) will now be used together with the double limit property to find step by step a sequence of functions which converges to the given

$$f_1: = f \in \overline{A}^{Z^X} .$$

By (a) (and the fact that $f|_{\overline{D}_1}$ is continuous), there is a finite set $L_1^1 \subset D_1$ with

$$\min_{y\in L_1^1} d(f_1(x),f_1(y)) \le 1 \quad \text{for all} \quad x \in \overline{D}_1 .$$

But f is in the closure of A, so there is an $f_2 \in A$ with

$$\max_{y\in L_1^1} d(f_2(y),f(y)) \le \frac{1}{2} .$$

Proceeding by induction there are finite sets $L_n^i \subset D_i$, $i \le n$, and functions $f_n \in A$ such that

$$\min_{y\in L_n^i}\max_{k\le n} d(f_k(x),f_k(y)) < \frac{1}{n} \quad \text{for all} \quad x\in \overline{D}_i$$

and

$$\max_{y\in\bigcup\{L_j^i \mid i\le j\le n\}} d(f_{n+1}(y),f(y)) < \frac{1}{n+1}$$

(c) To show that for $x\in\overline{D}_i$ $\lim_{n\to\infty} f_n(x) = f(x)$ in the compact space Z it is enough that the only cluster-point of $(f_n(x))$ is $f(x)$:
By (b) there are $y_n \in L_n^i \subset D_i$ with

$$(*) \qquad \max_{k\le n} d(f_k(x),f_k(y_n)) < \frac{1}{n}$$

and

$$\lim_m f_m(y_n) = f(y_n) .$$

For $k = 1$ (*) implies therefore

$$\lim_n \lim_m f_m(y_n) = f_1(x) = f(x)$$

and for $k = m$

$$\lim_n f_m(y_n) = f_m(x) .$$

The interchangeable double-limit-property together with the fact that Z is compact and metric ascertains that $f(x)$ is the only cluster-point of $(f_n(x))$ and therefore $\lim_n f_n(x) = f(x)$ for all $x\in X$. ∎

3.7. After these preparations, a large class of spaces $C(X,Z)$ can be diagnosed as being angelic. The following main theorem of this paragraph constitutes a combination of results which were obtained by A. Grothendieck, D.H. Fremlin, J.D. Pryce, and M. De Wilde.

THEOREM: If $X = \overline{\bigcup_{n=1}^{\infty} K_n}$, K_n relatively countably compact and Z is metric, then C(X,Z) is ω_X-angelic.

Proof:

(a) Since

$$C(X,\mathbb{R}) \hookrightarrow C(X,\overline{\mathbb{R}})$$

the angelic-theorem 3.3, and Fremlin's theorem 3.5, imply, that Z can be assumed metric and compact. Furthermore, again 3.3, applied to

$$C(X,Z) \hookrightarrow C(\biguplus_{n=1}^{\infty} \overline{K}_n, Z)$$

(disjoint union) allows to consider only spaces of the form

$$X = \biguplus_{n=1}^{\infty} \overline{K_n}$$

K_n relatively countably compact.

(b) In this case, take an ω_X-relatively countably compact subset $A \subset C(X,Z)$ and

$$f \in \overline{A}^{Z^X} ,$$

then

$$f|_{\overline{K}_n} \in \overline{A|_{\overline{K}_n}}^{Z^{\overline{K}_n}} \subset C(\overline{K}_n, Z)$$

by theorem 1.4. and, according to the special nature of X , f is continuous: this shows, that A is ω_X-relatively compact. The more, De Wilde's theorem 3.5. tells that f is the limit of a sequence in A ; both characteristics of angelic spaces are therefore satisfied. ∎

The theorem holds in particular for separable spaces X , for locally compact spaces which are countable at infinity, $\sigma(E',E)$-duals of normed spaces (this will be improved).

Observe that by 3.4. the assumption "Z metric" can be replaced by: Z is regular and admits a continuous metric.

3.8. The theorem states that in many spaces $C(X,Z)$ much of the pointwise topology is actually determined by countable sets. The following result reinforces this aspect:

THEOREM (I.Kaplansky): Let $X = \bigcup_{n=1}^{\infty} K_n$, K_n compact, Z metric, and $A \subset C(X,Z)$. Then for every $f_o \in \overline{A}^X$ (in $C(X,Z)$) there is a countable subset $D \subset A$ with

$$f_o \in \overline{D}^{\omega_X} .$$

Proof: It is evident, that the theorem needs only to be proved for compact X. In this case a countable $D \subset A$ is demanded such that

$$\forall_m \forall_n \forall_{x_1,\dots,x_n \in X} \exists_{g \in D} \forall_{i=1,\dots,n} \; d(f_o(x_i),g(x_i)) < \frac{1}{m}$$

Fix m and n and define for every $g \in A$ the open set

$$L_g^{m,n} := \{(x_1,\dots,x_n) \in X^n \mid d(f_o(x_i),g(x_i)) < \frac{1}{m}\};$$

$f_o \in \overline{A}^{\omega_X}$ says that

$$X^n \subset \bigcup_{g \in A} L_g^{m,n}$$

is an open cover of the compact space X^n. Hence there is a finite set $D_{m,n} \subset A$ with

$$X^n \subset \bigcup_{g \in D_{m,n}} L_g^{m,n} :$$

$D := \bigcup_{m,n} D_{m,n}$ satisfies the theorem. ■

In the light of the foregoing results it might be astonishing that Kaplansky's theorem is false for countably compact sets X: for an example, consider (see 2.1). the countably compact ordinal space $W := [1,\omega_1[$, the unit interval $I := [0,1]$,

$$A := \{f \in C(W,I) \mid \lim_{x\to\omega_1} f(x) = 0\}$$

and note that $1 \in \bar{A}$ since W is completely regular, but every function in A has countable support.

The Kaplansky theorem can be used to prove the main theorem 3.7. for spaces $X = \overline{\cup K_n}$, K_n compact (see Ex. 3.20.).

3.9. If E is a locally convex space, then certainly

$$E \subset C((E',\sigma(E',E)),\mathbb{R})$$

and $\omega_{E'}$ induces $\sigma(E,E')$ on E. Therefore it is easy to pass $C(X,Z)$-theorems to locally convex spaces.

First, if E is metrizable, (U_n) a basis of zero-neighbourhoods, then

$$E' = \bigcup_{n=1}^{\infty} U_n^o ,$$

the U_n^o being $\sigma(E',E)$-compact by the theorem of Alaoglu-Bourbaki, hence Kaplansky's theorem yields

COROLLARY: If E is a metrizable locally convex space, $A \subset E$, then for every $x \in \bar{A}^{\sigma(E,E')}$ there is a countable $D \subset A$ with

$$x \in \bar{D}^{\sigma(E,E')} .$$

3.10. Subspaces of angelic spaces are angelic; so the main-theorem 3.7. implies the

THEOREM (W.F.Eberlein-V.L.Šmulian): A locally convex space E which admits $\sigma(E',E)$-relatively countably compact sets $K_n \subset E'$, such that

$$(*) \qquad E' = \overline{\bigcup_{n=1}^{\infty} K_n}^{\sigma(E',E)}$$

is $\sigma(E,E')$-angelic.

- that is, all the statements of Šmulian's theorem 3.2. hold. Šmulian's theorem (the assumption was that E' is $\sigma(E',E)$-separable) is certainly a special case. But much more: the argument in 3.9. gives

(1) All metrizable locally convex spaces, in particular all normed spaces, are angelic in their weak topology.

Observe that, if E is normed, E' is $\sigma(E',E'')$-angelic - but not necessarily $\sigma(E',E)$-angelic (see 1.2.(5) and Ex. 3.22. for examples).*)

The angelic-theorem 3.3.(2) says that a space is angelic if there is a coarser angelic topology on it: if $T: E \to F$ is linear, continuous (E,F locally convex), then T is continuous with respect to $\sigma(E,E')$ and $\sigma(F,F')$, hence

(2) If there is a coarser metrizable locally convex topology on a locally convex space (E,τ), then (E,τ) is angelic in its weak topology.

(This, as well as the result (4) for strict (LF)-spaces is due to J. Dieudonné and L. Schwartz.)
The assumption is satisfied, if there is simply a continuous norm! Furthermore,

(3) All sequence-spaces (i.e. vector-subspaces of $\mathbb{K}^{\mathbb{N}}$ with a locally convex topology finer than the pointwise convergence) are angelic in their weak topology.

Those classes of (LF)-spaces which usually appear in applications behave also well.

(4) Strict (LF)-spaces and (LF)-spaces with weakly compact linking mappings are angelic in their weak topologies.

Proof: Both classes of spaces have the property that bounded sets $A \subset E = \text{ind}\, E_n$ are already bounded in some E_m. So take a $\sigma(E,E')$-relatively countably compact set A:
(a) If E is a strict (LF)-space, E_m is (with its Fréchet-space topology and therefore also for the weak topology) a closed subspace of E: but by (1) E_m is angelic in its weak topology.

*)For a smooth Banach-space E, the dual unit ball is $\sigma(E',E)$-sequentially compact; see a forthcoming paper by J.Hagler and F.Sullivan (Notices 1979(1)763-46-21)).

(b) If the linking mappings are weakly compact, then $A \subset \lambda U_m$, where U_m is a neighbourhood of zero in E_m which is contained in a $\sigma(E_{m+1}, E'_{m+1})$-compact subset $B \subset E_{m+1}$: on λB the weak topologies of E_{m+1} and E coincide. ■

A particular case is when the linking mappings are compact as in spaces of germs of holomorphic functions. The space $\mathcal{D}(\Omega)$ of test-functions of distributions on an open set $\Omega \subset \mathbb{R}^n$ is a strict (LF)-space, as well as the space $C_{oo}(X)$, X locally compact and countable at infinity, of continuous functions with compact support.

The inductive limit of a sequence of reflexive Banach-spaces has certainly weakly compact linking mappings; however, a result of W. J. Davis, T. Figiel, W. B. Johnson, and A. Pełczyński tells that all weakly compact mappings factor through a reflexive Banach-space.

For spaces of operators the main-theorem 3.7. directly applies via $L(E,F) \subset C(E,F)$. Remembering, that the topology in F can always be refined (3.4.), yields

> (5) If $E = \overline{\bigcup_{n=1}^{\infty} K_n}$, K_n countably compact, is a locally convex space, F a locally convex space with a coarser metrizable topology, then the space $L(E,F)$ of linear continuous operators is angelic in the topology of pointwise convergence.

Take for example F any locally convex space with a continuous norm and $\mathcal{D}(\Omega, F) := L(\mathcal{D}(\Omega), F)$ the space of F-valued distributions, with the topology of pointwise convergence.

3.11. It might be redundant to mention again that, by the angelic theorem 3.3.(2), also all finer regular topologies are angelic - in spaces of operators (under the assumption of (5)) the topology of uniform convergence on any covering of E by bounded sets ...

COROLLARY 1: For a locally convex space with (*) p. 38 (or any of the spaces of 3.10.(1)-(4)) all (regular) topologies on E which are finer than $\sigma(E,E')$ are angelic.

Note, that separable locally convex spaces need not be angelic (1.2.(6))!

3.12. The following characterization of semi-reflexive spaces is immediate:

COROLLARY 2: For a locally convex space E with (*) (or any of the spaces of 3.10.(1)-(4)) the following are equivalent:

(a) E is semi-reflexive.

(b) Every bounded sequence has a $\sigma(E,E')$ - convergent subsequence.

Of different type is the

COROLLARY 3: A $\sigma(E,E')$-separable complete locally convex space E is semi-reflexive if and only if every $\sigma(E',E)$-convergent sequence in E' is $\sigma(E',E'')$-convergent.

According to a theorem of J. Dieudonné and A. P. Gomes a separable Fréchet-space is Montel if the $\sigma(E',E)$-convergent sequences in E' converge strongly.

Proof: The Eberlein-Šmulian-theorem implies, that $(E',\sigma(E',E))$ is angelic. Therefore every polar U^o of a zero -neighbourhood $U \in \mathfrak{U}_E(0)$ is $\sigma(E',E)$-sequentially compact. The assumption implies now, that U^o is $\sigma(E',E'')$-countably compact, and hence, by the angelic-lemma, the (continuous) identity-map

$$(U^o,\sigma(E',E'')) \to (U^o,\sigma(E',E))$$

is a homeomorphism. Therefore, every $x \in E''$ has $\sigma(E',E)$-continuous restrictions to equicontinuous subsets of E' and Grothendieck's completeness criterion (0.6.) implies $E'' = E$. ■

EXERCISES

3.13. Is Tychonoff's Plank (2.1) angelic?

3.14. If $C(X,Z)$ is ω_X-angelic then Z is angelic.

3.15. (a) Complete locally convex spaces are not always angelic in their weak topology.

(b) $C(X,\mathbb{R})$ is not always ω_X-angelic.

(c) More precisely: Find an example of a topological space S and a sequentially compact subset $A \subset C(S,\mathbb{R})$ (with the ω_X-topology), whose closure is not countably compact. (Hint: Use the locally convex space from 1.2.(9), $E' = \text{span } X \subset E^*$ via Dirac-functionals, (1.24) and note, that χ_X is not $\sigma(E',E)$-continuous on $S := (E',\sigma(E',E))$.)

3.16. Are compact spaces angelic?

3.17. The arbitrary product of angelic spaces is not angelic. (T. K. Boehme and M. Rosenfeld constructed two compact angelic spaces, whose product is not angelic. W. Govaerts found a subclass of the class of angelic spaces which satisfies 3.3., 3.5., 3.7., and is closed under taking countable products.*) The additional defining condition is: In every separable compact subset each point is the intersection of countably many open sets.)

3.18. De Wilde's theorem 3.6. does not hold if only $X = \overline{\bigcup D_m}$ is supposed, D_m countably compact. (Hint: $\mathbb{R} = \overline{\mathbb{Q}}$ and functions of the form $x \rightsquigarrow \min(1, |x - x_o|)$.)

3.19. Assume $D \subset X$ is dense, $A \subset C(X,\mathbb{R})$ and $A \sim D$ (in $\overline{\mathbb{R}}$),

(a) A is ω_X-angelic (Hint: 1.5.(2) and De Wilde's theorem)

(b) A is ω_D-angelic

(c) A sequence (f_n) in A ω_D-converges if and only if it ω_X-converges

3.20. (a) If K is countably compact, Z metric, $B \subset C(K,Z)$ countable, then there is a set $K/\!/B$ and an onto map

$$\kappa : K \longrightarrow K/\!/B$$

*) There is another remarkable subclass which enjoys this property: Those compact sets which are homeomorphic (equivalently: continuous) images of weakly compact subsets of Banach-spaces (= Eberlein-compacta). See a forthcoming paper of Y.Benyamini, M.E.Rudin, and M.Wage.

such that the quotient topology on $K//B$ is compact and metrizable and every $f \in B$ factors through κ. (Hint: Consider

$$\Phi : K \longrightarrow Z^{B}$$
$$x \rightsquigarrow (f(x))_{f \in B}$$

and apply the angelic lemma on the two topologies of $K//B := \Phi(K)$.)

(b) If $X = \overline{\bigcup_{n=1}^{\infty} K_n}$, K_n countably compact, Z metric, then for every ω_X-separable, ω_X-relatively countably compact subset $A \subset C(X,Z)$ the ω_X-closure $\bar{A}$ is metrizable. (Hint: Study the diagram

$$\begin{array}{ccc} \bar{A} & \hookrightarrow & C(X,Z) \\ \downarrow & & \downarrow \\ \prod_n C(K_n//A,Z) & \hookrightarrow & \prod_n C(K_n,Z) \\ \downarrow & & \\ Z^{\mathbb{N} \times \mathbb{N}} & & \end{array}$$

.)

This result, together with Kaplansky's theorem provides a proof of the main theorem 3.7. in a slightly weaker form.

3.21. If for locally convex spaces $E \neq \{0\}$ and F the space of operators $L(E,F)$ is angelic in, say, the topology of uniform convergence on all bounded sets, then F is angelic. (Hint: $F \ni y \rightsquigarrow \varphi \otimes y \in L(E,F)$)

3.22. Let (X, Ω, μ) be a measure space such that $(L^1)' = L^\infty$.

(a) If L^1 is norm-separable (this is true, e.g., for Radon-measures on compact metrizable spaces), then $\sigma(L^\infty, L^1)$ is angelic.

(b) If there is a non-σ-finite measurable set, then $(L^\infty, \sigma(L^\infty, L^1))$ is not angelic. (Hint: Use that the support of an integrable function is σ-finite)

3.23. (a) Let E and F be metrizable locally convex spaces and $f: E \to F$ a (not necessarily linear) map. If f is "completely continuous", i.e.: for every $\sigma(E,E')$-convergent sequence (x_n) in E the sequence $(f(x_n))$ converges, then f is continuous and maps weakly compact sets onto compact sets.

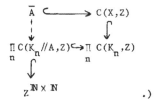

(b) If, additionally, E is reflexive, f maps bounded sets onto relatively compact sets.

3.24. Show that for a separable locally convex space E every $\beta(E',E)$ - relatively countably compact sequence in E has a $\beta(E',E)$-convergent subsequence.

3.25. Every bounded sequence in a Hilbert-space has a weakly convergent subsequence.

3.26. A sequentially continuous function $f: X \to Y$, X angelic and compact, is continuous. (Hint: Consider $f^{-1}(A)$, A closed.)

3.27. Let E be separable (or more general: containing a sequence of $\sigma(E,E')$-relatively countably compact sets with dense union), complete locally convex space.

(a) Every $x \in E'^*$ such that $\langle x,\varphi_n\rangle \to 0$ for all equicontinuous $\sigma(E',E)$-zero-sequences, is in E. (Hint: 3.26. and Grothendieck's completeness-theorem.)

(b) Let (X, Ω, μ) be a measure space and assume for a weakly integrable function $f: X \to E$ (i.e. $\langle\varphi,f(\cdot)\rangle$ is integrable for all $\varphi \in E'$) that

$$\lim_n \int_X \langle\varphi_n,f(t)\rangle\ \mu(dt) = 0$$

for all equicontinuous $\sigma(E',E)$-zero-sequences (φ_n). Then there is an $x \in E$ with

$$\int_X \langle\varphi,f(t)\rangle\ \mu(dt) = \langle\varphi,x\rangle$$

for all $\varphi \in E'$.

These results are due to C. Constantinescu.

3.28. Deduce directly from Šmulian's theorem 3.2. that in a locally convex space E with $E' = \overline{\cup K_n}^{\sigma(E',E)}$, K_n $\sigma(E',E)$-relatively countably compact, every $\sigma(E,E')$-(relatively) countably compact set is $\sigma(E,E')$-(relatively) sequentially compact. (Hint: Assume, which is enough, that E is $\sigma(E,E')$-separable; then $\overline{K_n}$ is metrizable, compact therefore separable.)

§4. POINTWISE AND WEAK COMPACTNESS IN SPACES OF CONTINUOUS FUNCTIONS

4.1. If for a topological (Hausdorff) space X the space $C(X) := C(X,\mathbb{R})$ of continuous functions is ω_X-angelic, then it is angelic for all finer (regular) topologies such as

(1) the <u>compact-open-topology</u> $\tau_{co}(X)$, that is the topology of uniform convergence on all compact sets,

(2) the topology $\tau_{bdg}(X)$ of <u>uniform convergence on all bounding (see 2.1.) subsets</u> of X.

Both, τ_{co} and τ_{bdg}, are locally convex (Hausdorff) topologies defined by the semi-norms

$$p_K(f) := \sup_{x \in K} |f(x)|$$

K compact or bounding respectively. Therefore

(3) the weak topology $\sigma_{co}(X)$ associated to $\tau_{co}(X)$, and

(4) the weak topology $\sigma_{bdg}(X)$ associated to $\tau_{bdg}(X)$

both being finer than ω_X, are also angelic.

The space $C^{b}(X)$ of bounded continuous functions is normed, so the associated weak topology is always angelic by the Eberlein-Šmulian-theorem 3.10.(1).

4.2. Therefore if K is compact, ω_K (by 3.7.) and $\sigma_{co}(K)$ are both angelic but may have even different convergent sequences (Ex. 4.10.) and therefore different compact sets. However, the following result of A. Grothendieck holds

THEOREM: If K is compact then a subset $A \subset C(K)$ is weakly compact if and only if it is uniformly bounded and ω_K-compact.

("Weakly" refers to $\sigma_{co}(K)$, the weak topology of the normed spaces $C(K)$.)

For convex sets A the condition of uniform boundedness is redundant (see 4.9.).

Proof: The weak topology is finer than ω_K and, by Mackey's theorem, every weakly bounded set is bounded in $C(K)$, i.e. uniformly bounded.

On the other hand, by the fact that both topologies are angelic, it is enough to show, that every uniformly bounded pointwise convergent sequence is weakly convergent: But, observing that every $\varphi \in C(K)'$ is represented by a measure this is true by Lebesgue's dominated convergence theorem. ∎

4.3. For arbitrary topological (Hausdorff)-spaces X, the initial topologies of the embedding

$$\Psi : C(X) \hookrightarrow \prod_{\substack{K \subset X \\ \text{compact}}} C(K)$$

$$f \rightsquigarrow (f|_K)_K$$

are just the right topologies, namely

(a) $\Pi\,\omega_K$ induces ω_X on $C(X)$

(b) $\Pi\,\tau_{co}(K)$ induces $\tau_{co}(X)$ on $C(X)$

and, according to the behavior of weak locally convex topologies with respect to products and subspaces, also

(c) $\Pi\,\sigma_{co}(K)$ induces $\sigma_{co}(X)$ on $C(X)$.

Recall that ω_K and $\sigma_{co}(K)$ are angelic.

COROLLARY 1: For a ω_X-relatively countably compact set $A \subset C(X)$, which is $\tau_{co}(X)$-bounded (that is uniformly bounded on all compact sets $K \subset X$) the topologies $\sigma_{co}(X)$ and ω_X coincide on $\overline{A}^{\omega_X}$; in particular: the weak and pointwise closure of A are the same.

Proof: $A|_K$ is ω_K-relatively (countably) compact in $C(K)$ and uniformly bounded. The theorem says that

$$B_K := \overline{A|_K}^{\omega_K}$$

is $\sigma_{co}(K)$-compact, and therefore ω_K and $\sigma_{co}(K)$ coincide on B_K as well as $\Pi\omega_K$ and $\Pi\sigma_{co}(K)$ are the same on the (compact) set $\Pi B_K \supset A$. ∎

A need not be ω_X-relatively compact (see Ex. 4.12.(a)) but Corollary 1 immediately implies the following result of C. Constantinescu (for convex sets see 4.9.):

COROLLARY 2: Let $A \subset C(X)$ be uniformly bounded on all compact subsets of X . Then

(1) A is ω_X-compact (resp. countably compact, sequentially compact, relatively ...) if and only if it is $\sigma_{co}(X)$-compact (resp. countably compact, ...)

(2) Each sequence (f_n) in A converges pointwise in X if and only if it converges weakly (i.e. with respect to $\sigma_{co}(X)$) , the same holds for Cauchy-sequences.

The result for Cauchy-sequences follows immediately from the observation (Ex. 4.25.) that a sequence (f_n) is Cauchy if and only if for all subsequences (g_m) of (f_n) the sequence $(g_m - g_{m+1})$ converges to zero.

The statement (2) is a kind of Lebesgue's dominated convergence theorem.

4.4. Investigating the normed space $C^b(X)$ of bounded continuous functions on a completely regular space X it is natural to look at the Stone-Čech-compactification βX of X : One way to obtain it, is to embed

$$\begin{array}{rcl} S: X & \hookrightarrow & \mathbb{R}^{C^b(X)} \\ \cup & & \cup \\ x & \rightsquigarrow & (f(x))_f \end{array} \quad ;$$

since X is completely regular, it is easily checked that the product-topology induces the original topology on X . $S(X)$ is coordinate-wise bounded, hence it is relatively compact by the Tychonoff-theorem - and $\beta X := \overline{SX}$ has the property that every continuous function $f \in C^b(X)$ has a unique extension $f^\beta \in C(\beta X)$-namely the projection on the f-coordinate. By a simple compactness argument

$$f^\beta(\beta X) = \overline{f(X)} \quad .$$

Therefore

$$\sup_{x \in X} |f(x)| = \sup_{x \in \beta X} |f^\beta(x)|$$

and $C^{b}(X) = C(\beta X)$ with the same norms: Hence the weak topologies coincide. "The" weak topology as well as $\omega_{\beta X}$ are angelic (4.2) but ω_X need not be (Ex. 4.12.(b)).

COROLLARY 3: For a subset $A \subset C^{b}(X)$, X completely regular, the following are equivalent:

(1) A is weakly relatively compact

(2) A is uniformly bounded on X and $\omega_{\beta X}$-relatively compact

(3) A is uniformly bounded on X and has interchangeable double-limits with X: $A \sim X$.

Proof (1)⟺(2) by the theorem, (2)⟺(3) by 1.4. ∎

Certainly (1)-(3) imply that A is ω_X- relatively compact. But there may be uniformly bounded ω_X-convergent sequences which do not $\omega_{\beta X}$-converge (see Ex. 4.13. for an example).

4.5. There is an analogous "completion" of a completely regular space X which allows to extend all continuous functions and helps to describe the topology $\tau_{bdg}(X)$ of uniform convergence on all bounding subsets of X: The repletion (real-completion, real-compactification) of a completely regular (Hausdorff) space is the set

$$\upsilon X := \{\varphi: C(X) \to \mathbb{R} \mid \varphi \text{ linear, multiplicative, } \varphi(1) = 1\} \subset \mathbb{R}^{C(X)}$$

of all characters on $C(X)$, endowed with the product topology; υX is completely regular and a closed subspace of $\mathbb{R}^{C(X)}$. Via the Dirac-functionals

$$\delta: X \hookrightarrow \upsilon X$$
$$x \rightsquigarrow \delta_x \quad ; \quad \delta_x(f) := f(x)$$

X is a subset of υX - even in its topology, since X is completely regular. X is called replete (real complete, real compact) if δ is onto: $X = \upsilon X$. The repletion of a space, compact spaces and separable metric spaces are all replete (see Ex. 4.14., 4.18., 4.19.). (The c-repletion due to H. Buchwalter can as well be used to study $\sigma_{bdg}(X)$.)

Every $f \in C(X)$ has a natural extension $f^{\upsilon} \in C(\upsilon X)$: the restriction of the f-projection of $\mathbb{R}^{C(X)}$ to υX

$$f^{\upsilon}(\varphi) := \varphi(f) \qquad f \in C(X),\ \varphi \in \upsilon X \ .$$

THEOREM: $$f^{\upsilon}(\upsilon X) = f(X) \ .$$

Proof: Assume, for fixed $f \in C(X)$, $\varphi \in \upsilon X$, that

$$\alpha := \varphi(f) = f^{\upsilon}(\varphi) \notin f(X) \ ;$$

then the function $(f - \alpha)^{-1}$ exists and is continuous: But

$$1 = \varphi(1) = \varphi((f-\alpha)^{-1}(f-\alpha)) = \varphi((f-\alpha)^{-1}) \cdot (\varphi(f) - \alpha) = 0 \ . \qquad \blacksquare$$

In other words (see Ex. 4.15.): The range of f^{υ} is the spectrum of f in the algebra $C(X)$.

4.6. The theorem says, that for every $\varphi \in \upsilon X$ and every $f \in C(X)$, there is an $x \in X$ such that

$$f^{\upsilon}(\varphi) = f(x) \ .$$

However, much more is true:

(1) For every $\varphi \in \upsilon X$ and every countable set $A \subset C(X)$ there is an $x \in X$ with

$$f^{\upsilon}(\varphi) = f(x)$$

for all $f \in A$.

Proof: For $A = \{f_n\}$ the sum

$$g := \sum_{n=1}^{\infty} 2^{-n} \frac{(f_n - \varphi(f_n))^2}{1 + (f_n - \varphi(f_n))^2} \qquad \in C(X)$$

is uniformly convergent on X

$$\sup_{x\in X} (g - \sum_1^n)^2 =: \epsilon_n \to 0$$

and therefore, by the theorem,

$$(\varphi(g))^2 = (\varphi(g) - \varphi(\sum_1^n))^2 = \varphi((g - \sum_1^n)^2) = ((g - \sum_1^n)^2)^{\upsilon}(\varphi) \leq \epsilon_n$$

that is $g^{\upsilon}(\varphi) = \varphi(g) = 0$. Again by the theorem, there is an $x \in X$ with

$$0 = g^{\upsilon}(\varphi) = g(x) = \sum_1^{\infty} \ldots$$

and therefore all the terms in the sum have to be zero:

$$f_n(x) = \varphi(f_n) \ . \qquad \blacksquare$$

In particular: For every $\varphi \in \upsilon X$, every $f_1, \ldots, f_n \in C(X)$ and every $\epsilon > 0$, there is an $x \in X$ with

$$\max_i | f_i(x) - f_i^{\upsilon}(\varphi) | = 0 \leq \epsilon$$

that is

(2) X is dense in υX, $\overline{X}^{\mathbb{R}^{C(X)}} = \upsilon X$ and the extension f^{υ} of an $f \in C(X)$ is the unique $g \in C(\upsilon X)$ with $g|_X = f$.

In this sense $C(X) = C(\upsilon X)$ (as algebras, see ex. 4.15.(2)). If X is not replete, the topology $\omega_{\upsilon X}$ is strictly finer than ω_X, since a $\varphi \in \upsilon X \setminus X$ is certainly not ω_X-continuous. However, it is immediate by (1), that

(3) The map

$$C(X) \longrightarrow C(\upsilon X)$$
$$f \rightsquigarrow f^{\upsilon}$$

is ω_X - $\omega_{\upsilon X}$ - continuous on countable subsets of $C(X)$, in particular it is sequentially continuous.

Therefore

(4) A subset $A \subset C(X)$ is ω_X-countably compact if and only if it is $\omega_{\upsilon X}$-countably compact. The same holds true for sequentially compact, relatively countably compact, and relatively sequentially compact.

4.7. The following characterization of bounding sets is crucial

PROPOSITION: A subset of X is bounding if and only if it is relatively compact in the repletion υX.

Proof: The condition is clearly sufficient. Take therefore a bounding subset $Y \subset X$, that is all $f(Y) \subset \mathbb{R}$ are relatively compact.

$$Y \subset \prod_{f \in C(X)} \overline{f(Y)} \subset \mathbb{R}^{C(X)}$$

hence Y is relatively compact in $\mathbb{R}^{C(X)}$. But υX is closed in $\mathbb{R}^{C(X)}$ and so Y is also relatively compact in υX. ■

4.8. Now, the situation is nearly as in 4.3.: Since

$$\sup_{y \in Y} |f(y)| = \sup_{y \in \overline{Y}} |f^{\upsilon}(y)| ,$$

the map

$$C(X) \hookrightarrow \prod_{\substack{Y \subset X \\ \text{bounding}}} C(\overline{Y})$$

(the closure of Y taken in υX) induces on $C(X)$

(a) the topology $\tau_{bdg}(X)$ of uniform convergence on all bounding subsets of X, if the spaces $C(\overline{Y})$ carry their norm-topology.

(b) the associated weak topology $\sigma_{bdg}(X)$, if all $C(\overline{Y})$ carry their weak-topology.

- but the product of the pointwise topologies may be strictly finer than ω_X: A character

$$\varphi \in \Big(\bigcup_{\substack{Y \subset X \\ \text{bounding}}} \overline{Y} \Big) \setminus X$$

is not ω_X-continuous.

However 4.6.(4) stated, that ω_X- and $\omega_{\upsilon X}$- relatively countably compact subsets are the same. This is sufficient to prove the

THEOREM: Let X be a completely regular space and $A \subset C(X)$ uniformly bounded on all bounding subsets of X :

(1) If A is ω_X- relatively countably compact, then the topologies ω_X and $\sigma_{bdg}(X)$ coincide on $\overline{A}^{\omega_X}$.

(2) A is $\sigma_{bdg}(X)$-compact (resp. countably compact, sequentially compact, relatively ...) if and only if it is ω_X-compact (resp. countably ...)

(3) Each sequence (f_n) in A converges pointwise in X if and only if it converges with respect to $\sigma_{bdg}(X)$, the same for Cauchy-sequences.

As for the compact-open-topology, A as in (1) need not be ω_X-relatively compact (Ex. 4.12.). For convex sets A see 4.9. .

Proof: (2) and (3) are consequences of (1) and Ex. 4.25. .

To see (1), note that by 4.6.(4) A is $\omega_{\upsilon X}$ - and therefore $\omega_{\overline{Y}}$ -relatively countably compact for all bounding $Y \subset X$. But now, as in the proof of Corollary 1 in 4.3., Grothendieck's result 4.2. on the topologies on $C(K)$, K compact, says, that $B_Y := \overline{A|_{\overline{Y}}}$ is weakly compact and the weak and pointwise topologies $\omega_{\overline{Y}}$ and σ_Y of $C(\overline{Y})$ coincide on B_Y . Since

$$\overline{A}^{\omega_X} \subset \prod_Y B_Y$$

and $\Pi\omega_Y$ induces ω_X , (b) implies (1) . ∎

These results are due to J. Schmets and M. De Wilde.

4.9. For convex subsets of $C(X)$ the assumption of being uniformly bounded on all bounding (or compact) subsets of X in 4.8. (or 4.3. as well as in Corollary 3 (2) in 4.4.) is redundant:

PROPOSITION: If $A \subset C(X)$ is convex and ω_X-relatively countably compact, then A is uniformly bounded on all bounding subsets $Y \subset X$.

Proof: The set

$$B := \{f \in C(X) \mid \sup_{x \in Y} |f(x)| \leq 1\}$$

is ω_X-closed, absolutely convex and absorbing, therefore an ω_X - barrel: Lemma 1.6. states that $A \subset \lambda B$ for some $\lambda > 0$, i.e. A is uniformly bounded on Y . ∎

The following result is essentially due to I. Tweddle:

COROLLARY: If X is a completely regular Hausdorff space, admitting countably many bounding subsets X_n whose union is dense, then

(1) every convex ω_X-(relatively) countably compact set is (relatively) compact and (relatively) sequentially compact with respect to $\sigma_{co}(X)$.

(2) in particular, $\sigma_{co}(X)$, $\sigma_{co}(\upsilon X)$, $\sigma_{bdg}(X)$, ω_X , and $\omega_{\upsilon X}$ have the same convex compact (= countably compact = sequentially compact, relatively ...) sets.

Proof: If A is ω_X-(relatively) countably compact, it is $\omega_{\upsilon X}$-(relatively) countably compact by 4.6.(4). All X_n are relatively compact in υX (4.7.)

therefore (3.7.) $C(\upsilon X)$ is $\omega_{\upsilon X}$-angelic and A is $\omega_{\upsilon X}$-(relatively) compact and $\omega_{\upsilon X}$-(relatively) sequentially compact. If A is additionally convex, the proposition ascertains that A is $\sigma_{co}(\upsilon X)$-bounded and the Corollary 2 in 4.3. says that then A is $\sigma_{co}(\upsilon X)$-(relatively) compact and $\sigma_{co}(\upsilon X)$-(relatively)-sequentially compact. Consequently (2) is true, since ω_X is the weakest and $\sigma_{co}(\upsilon X)$ the strongest of the five topologies involved. ■

Note, that $C(X) = C(\upsilon X)$ was $\omega_{\upsilon X}$-angelic in this case.

EXERCISES

4.10. Give an example of a sequence (f_n) in $C([0,1])$ which converges pointwise but not weakly.

4.11. If X is locally compact, (f_n) a bounded sequence in $C_{co}(X)$ which converges pointwise to zero, then the closed absolutely convex hull of $\{f_n\}$ in $C_{co}(X)$ is $\sigma_{co}(X)$-compact. (Hint: Show that the map

$$\ell^1 \longrightarrow C(X)$$

$$(\xi_n) \rightsquigarrow \Sigma \xi_n f_n$$

is $\sigma(\ell^1, c_o)$-$\sigma_{co}(X)$-continuous.)

4.12. (a) Give an example of a topological space S and a $\tau_{bdg}(S)$-bounded ω_S-relatively countably subset $A \subset C(S)$ which is not ω_S-relatively compact. (Hint: Use 3.15(c), and note that the set A in 1.2.(9) is a Banach-disc which is therefore absorbed by every Y^o, $Y \subset S = E'$, which is $\sigma(E',E)$-bounded.)

(b) For an S as in (a) $C^{\ell}(S)$ is not ω_S-angelic. (Hint: Find an embedding $C(S) \to C^{\ell}(S)$.)

(c) $C^{\ell}(X)$ is ω_X-angelic if and only if $C(X)$ is.

4.13. (a) If $X \subset Y$ is dense, $K \subset X$ compact and $f,g\colon Y \to \mathbb{R}$ are continuous functions such that $f = g$ on $X \setminus K$, then $f = g$ on $Y \setminus K$.

(b) For $f_n(x) := \max(0, n(x-1)+1)$, $x \in [0,1[$, the extension f_n^{β} to the Stone-Čech-compactification $\beta[0,1[$ is 1 on $\beta[0,1[\setminus [0,1[$.

(c) (f_n) is uniformly bounded, $\omega_{[0,1[}$-convergent, but not $\omega_{\beta[0,1[}$-convergent in $C^b([0,1[)$.

4.14. Let K be compact

(a) Every character φ on $C(K)$ is represented by a Dirac-functional. (Hint: Find an open cover $U_1,\dots,U_n$ of K and $f_1,\dots,f_n \in C(K)$ such that $\varphi(f_i) \neq f_i(x)$ for all $x \in U_i$ and apply 4.6.(1)) .

(b) Compact spaces are replete.

In the following exercises X is always a completely regular space.

4.15. (a) Show that the equalities $C^b(X) = C(\beta X)$ and $C(X) = C(\upsilon X)$ are actually equalities of algebras.

(b) For $f \in C^b(X)$

$$f^{\beta}(\beta X) = \overline{f(X)} = \{\alpha \in \mathbb{R} \mid (f-\alpha) \text{ has no inverse in } C^b(X)\} .$$

(c) For $f \in C(X)$

$$f^{\upsilon}(\upsilon X) = f(X) = \{\alpha \in \mathbb{R} \mid (f-\alpha) \text{ has no inverse in } C(X)\} .$$

4.16. Show that $\beta X = \{\varphi : C^b(X) \to \mathbb{R} \mid \text{multiplicative, linear } \varphi(1) = 1\} \subset \mathbb{R}^{C^b(X)}$. (Hint: Use 4.14.(a) and 4.15.(a).)

4.17. $X \subset \upsilon X \subset \beta X$ as topological spaces. (Hint: Use 4.16. and the fact that the topology of υX is given by $C^b(\upsilon X) = C^b(X)$.)

4.18. The repletion is replete. (Hint: Use 4.15.(a).)

4.19. (a) X is replete, if $C(X)$ is ω_X-separable. (Hint: Use 4.6.(1) twice.)

(b) Separable metric spaces are replete; $\mathbb{R}$ is replete.

4.20. X is replete if and only if there is no superspace $Y \supset X$, such that X is dense in Y and all $f \in C(X)$ can be extended continuously to Y.

4.21. Every $f \in C(X,Y)$ has a unique extension $f^{\upsilon} \in C(\upsilon X, \upsilon Y)$.

4.22. (M. De Wilde) A subset $Y \subset X$ is bounding if and only if it has interchangeable double limits in $\overline{\mathbb{R}}$ with all ω_X-relatively (countably) compact subsets $A \subset C(X)$. (Hint: For the necessity use 4.6.(4), 4.7., and 1.4.)

4.23. The locally convex topologies $\omega_{\upsilon X}$ and ω_X on $C(X)$ admit the same bounded sets.

4.24. (S. Simons) If $C(X)$ is ω_X-angelic, $\mathcal{Q}$ a σ-algebra of subsets of X, $A \subset C(X)$ an ω_X-relatively compact subset of $\mathcal{Q}$-measurable functions, and (f_α) a pointwise convergent net of functions in A. Then $\lim f_\alpha$ is $\mathcal{Q}$-measurable. (Note that De Wilde's theorem 3.6. allows to ascertain measurability even for some non-continuous functions.)

4.25. In a locally convex space, a sequence (x_n) is a Cauchy-sequence if and only if for all subsequences (y_m) of (x_n) the sequence $(y_m - y_{m+1})$ converges to zero.

§5. BEST APPROXIMATIONS AND THE THEOREM OF R.C.JAMES

5.1. In 1933 S. Mazur observed that for a continuous linear functional φ on a (real) Banach-space E, $\|\varphi\| = 1$, the hyperplane $[\varphi = 1] := \{y \in E \mid \langle\varphi, y\rangle = 1$ has an element of least norm if and only if φ attains its supremum on the closed unit ball, that is the

LEMMA: For a normed space E, $\varphi \in E'$, $\|\varphi\| = 1$, the equalities

$$\inf_{\langle\varphi,y\rangle = 1} \|y\| = 1 = \sup_{\|x\|=1} \langle\varphi,x\rangle \quad \left[= \sup_{\|x\|\le 1} \langle\varphi,x\rangle\right]$$

hold.

Proof: The second equality is merely the definition of a norm. For the first take $y \in E$ with

$$1 = \langle\varphi,y\rangle \le \|\varphi\| \; \|y\| = \|y\|$$

therefore $\inf \|y\| \ge 1$.

On the other hand, again by the definition of the norm, for every $\varepsilon > 0$ there are $x_\varepsilon \in E$ with $\|x_\varepsilon\| = 1$ and $\langle\varphi, x_\varepsilon\rangle = 1 - \varepsilon$; consequently

$$\langle\varphi, \frac{x_\varepsilon}{1-\varepsilon}\rangle = 1, \qquad \left\|\frac{x_\varepsilon}{1-\varepsilon}\right\| = \frac{1}{1-\varepsilon} \xrightarrow[\varepsilon\to 0]{} 1$$

and $\inf \|y\| = 1$ is established. ■

5.2. In the theory of approximation the following terminology is used: let D be a subset of a metric space M

(1) An element $x_o \in M$ has a <u>best-approximation</u> $y_o \in D$ if

$$\inf_{y \in D} d(x_o, y) = d(x_o, y_o)$$

(2) D is called <u>proximinal</u>, if all $x \in M$ have a best approximation in D .

Now Mazur's result reads: 0 has a best approximation in $[\varphi = 1]$ if and only if φ attains its supremum on the closed unit ball . Moreover

PROPOSITION: For a normed space E , $\varphi \in E'$, $\|\varphi\| = 1$, the following statements are equivalent:

(1) φ attains its supremum on the closed unit ball

(2) For one (and then for every) $\alpha \in \mathbb{R}$ the hyperplane $[\varphi = \alpha]$ is proximinal.

Proof: For $\langle\varphi, x_o\rangle \neq \alpha$ the substitution

$$y = \frac{x_o - z}{\langle\varphi, x_o\rangle - \alpha}$$

yields

$$\inf_{\langle\varphi,z\rangle = \alpha} \|x_o - z\| = \inf_{\langle\varphi,y\rangle = 1} |\langle\varphi, x_o\rangle - \alpha| \, \|y\|$$

- that is, the existence of a best-approximation of x_o in $[\varphi = \alpha]$ is equivalent to the existence of a best-approximation of 0 in $[\varphi = 1]$. The lemma gives the result. ∎

5.3. Certainly, if the closed unit ball is weakly compact (equivalently: E is reflexive), every linear continuous = weakly continuous functional attains its supremum on it. But take for example

$$\varphi = (1 - \frac{1}{n}) \in \ell^\infty = (\ell^1)'$$

then there is no sequence $x = (\xi_n) \in \ell^1$ with

$$\sum_{n=1}^{\infty} |\xi_n| = 1 \qquad \text{and} \qquad \langle\varphi, x\rangle = \sum_{n=1}^{\infty} \xi_n (1 - \frac{1}{n}) = 1 \quad ,$$

therefore φ does not attain its supremum:

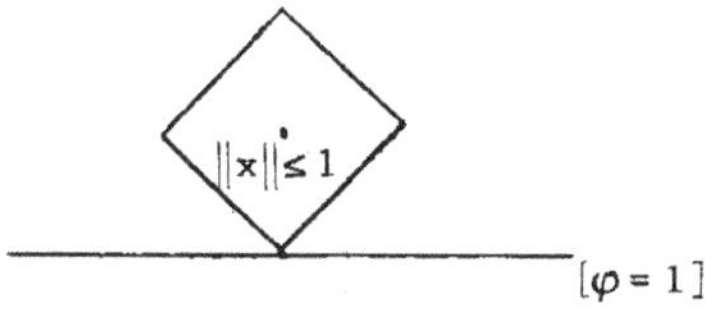

the closed unit ball lies on the hyperplane $[\varphi = 1]$ without touching it. There are many examples (see Ex. 5.7. - 5.11.) of this kind, so - does this happen in every non-reflexive Banach-space?

R.C. James announced in 1950 that a Banach-space with a basis is reflexive if it has the property that every continuous linear functional attains its supremum on the closed unit ball of all equivalent norms. In the same year, V. Klee proved this result without requiring a basis. Then, in 1957, R.C. James gave a sophisticated proof showing that the closed unit ball B of a separable Banach-space E is weakly compact (that is, E is reflexive) if (and only if) all $\varphi \in E'$ attain their supremum on B. In 1962, V. Klee conjectured that this property actually is characteristic for those weakly closed subsets of a (separable) Banach-space, which are weakly compact.

Finally in 1964, R.C. James published the following:

THEOREM (R.C.James): In a quasicomplete locally convex space E a weakly closed subset A is weakly compact if and only if all $\varphi \in E'$ attain their supremum on A.

The generalization to locally convex spaces was easy (see Ex. 5.13.). James' theorem is one of the deepest, most impressing, and most influential results in Functional Analysis. J.D. Pryce's version (1966) of R.C. James' proof will be given in § 6; nevertheless, free use of James' theorem will be made already in this paragraph and the exercises. It is immediate from the theorem that a Banach-space is reflexive, if and only if all $\varphi \in E'$ attain their supremum on the unit ball. The extension to locally convex spaces (Ex. 5.18.) was obtained by G. Köthe 1960 as a generalization of the before-mentioned result of V. Klee with the equivalent norms.

5.4. The translation of James' theorem into the language of approximation theory by use of proposition 5.2. is the

THEOREM: A Banach space is reflexive if and only if all closed hyperplanes are proximinal.

(There is a partial generalization to Fréchet - spaces, [19])

5.5. To extend this to all convex closed sets, the easy part of the following result, which is interesting for itself, will be helpful

THEOREM (J.Dieudonné-V.L.Šmulian): In a quasicomplete locally convex space a weakly closed set A is weakly compact if and only if for every sequence (C_n) of closed convex sets

$$\bigcap_{n=1}^{m} C_n \cap A \neq \emptyset \qquad m = 1, 2, \ldots$$

implies

$$\bigcap_{n=1}^{\infty} C_n \cap A \neq \emptyset \; .$$

Proof: Since for convex sets closed = weakly closed, the condition is clearly necessary.

Conversely, applying the James-theorem, take (for $A \neq \emptyset$) a $\varphi \in E'$,

$$\alpha := \sup_{x \in A} \langle \varphi, x \rangle \in \left]-\infty, +\infty\right]$$

and an increasing sequence (α_n) with $\lim_n \alpha_n = \alpha$ (in $\overline{\mathbb{R}}$) and $\alpha_n < \alpha$.

Then

$$C_n := \{x \in E \mid \langle \varphi, x \rangle \geq \alpha_n\}$$

satisfies

$$\bigcap_{n=1}^{m} C_n \cap A = C_m \cap A \neq \emptyset$$

and therefore

$$\{x \in A \mid \langle \varphi, x\rangle = \alpha\} = \bigcap_{n=1}^{\infty} C_n \cap A \neq \emptyset . \qquad \blacksquare$$

COROLLARY: For a Banach-space E the following statements are equivalent:

(1) E is reflexive.

(2) All closed (affine) hyperplanes are proximinal.

(3) All closed (affine) subspaces are proximinal.

(4) All closed, convex, non-empty subsets are proximinal.

(5) All weakly closed non-empty subsets are proximinal.

Proof: According to 5.4. only (1)⇒(5) needs a proof: If E is reflexive, A weakly closed, $x_0 \notin A$, and

$$d := \inf_{y \in A} \|x_0 - y\|$$

then the Dieudonné-Šmulian theorem ascertains that

$$\{y \in A \mid \|x_0 - y\| = d\} = \bigcap_{n=1}^{\infty} \{y \in A \mid \|x_0 - y\| \leq d + \tfrac{1}{n}\} \neq \emptyset . \qquad \blacksquare$$

5.6. James' theorem is not true in normed spaces, i.e. there are non-complete (and therefore non-reflexive) normed spaces, such that each $\varphi \in E'$ attains its supremum on the closed unit ball. The idea how to construct such a space, again due to R.C. James, is as follows: In a reflexive Banach space F , the (weakly compact) unit ball B is, by the Krein-Milman-theorem, the closed convex hull of its extreme points

$$B = \overline{\mathrm{co}\,\mathrm{ext}\,B}$$

therefore $E := \mathrm{span}\,\mathrm{ext}\,B$ is dense in F . Every $\varphi \in E' = F'$ attains its supremum in an extreme point of the unit ball of F , so on the unit ball of E (induced norm): If $E \neq F$, the desired counterexample is established.

The construction will be done by use of the

LEMMA: If $(F_n, \| \|_n)$ are Banach-spaces then

$$\ell^2(F_n) := \{(x_n) \in \prod_n F_n \mid \|(x_n)\| := (\sum_{n=1}^{\infty} (\|x_n\|_n)^2)^{\frac{1}{2}} < \infty\}$$

is also a Banach space. Its dual space is

$$(\ell^2(F_n))' = \ell^2(F_n')$$

(isometrically) with the duality bracket

$$\langle(\varphi_n),(x_n)\rangle = \sum_{n=1}^{\infty} \langle\varphi_n, x_n\rangle \quad .$$

In particular, if all F_n are reflexive spaces, then $\ell^2(F_n)$ is reflexive.

Proof: That $\ell^2(F_n)$ is a Banach-space is a routine-checking. For the dual space, first note

$$\Sigma \mid \langle\varphi_n, x_n\rangle \mid \le \Sigma \|\varphi_n\| \, \|x_n\| \le (\Sigma \|\varphi_n\|^2)^{\frac{1}{2}} \, (\Sigma \|x_n\|^2)^{\frac{1}{2}}$$

such that the duality-bracket is well-defined and

$$\|(\varphi_n)\|_{(\ell^2(F_n))'} \le (\Sigma \|\varphi_n\|^2)^{\frac{1}{2}} = \|(\varphi_n)\|_{\ell^2(F_n')} \quad .$$

On the other hand, define the natural injections

$$\iota_n : F_n \hookrightarrow \ell^2(F_m)$$

and, for a given $\varphi \in (\ell^2(F_m))'$, $\varphi_n := \varphi \circ \iota_n$; then

$$(*) \qquad \langle\varphi, x\rangle = \langle\varphi, \sum_n \iota_n x_n\rangle = \sum_n \langle\varphi, \iota_n x_n\rangle = \Sigma \langle\varphi_n , x_n\rangle \quad .$$

For an $\varepsilon > 0$, choose now $x_n \in F_n$, $\|x_n\| = 1$ with

$$\|\varphi_n\| = \langle \varphi_n , \frac{x_n}{1-\varepsilon} \rangle$$

then $(\xi_n x_n) \in \ell^2(F_n)$ for all $(\xi_n) \in \ell^2$ and, by (*),

$$|\sum_n \xi_n \|\varphi_n\| | = |\sum_n \xi_n \langle \varphi_n, \frac{x_n}{1-\varepsilon} \rangle| = |\langle \varphi, (\frac{\xi_n x_n}{1-\varepsilon})\rangle| \le$$

$$\le \|\varphi\| \frac{1}{1-\varepsilon} (\sum_n |\xi_n|^2)^{\frac{1}{2}}$$

therefore $(\sum \|\varphi_n\|^2)^{\frac{1}{2}} \le \|\varphi\|(1-\varepsilon)^{-1}$ for all $\varepsilon > 0$, $(\varphi_n) \in \ell^2(F'_n)$ and

$$(\ell^2(F_n))' = \ell^2(F'_n)$$

is isometrically. ∎

For the example, take $X_n := \{1,\dots,n\}$, the finite-dimensional $F_n := \ell^\infty(X_n)$, define

$$X := \bigcup_{n=1}^{\infty} X_n$$

(disjoint union) and write (ρ_n the restriction-mapping)

$$F := \ell^2(F_n) = \{f: X \to \mathbb{R} \mid (\sum \|\rho_n(f)\|_\infty^2)^{\frac{1}{2}} < \infty \} .$$

F is reflexive. If f is an extreme point of the unit ball B of F then $\rho_n(f)$ is an extreme point of the ball with radius $\|\rho_n(f)\|_\infty$ in $\ell^\infty(X_n)$, therefore

$$|\rho_n(f)(1)| = |\rho_n(f)(2)| = \dots = |\rho_n(f)(n)| .$$

The set

$$C: = \{f \in F \mid |\rho_n(f)(1)| = \dots = |\rho_n(f)(n)| \; n = 1,2,\dots\}$$

is certainly homogeneous and therefore

$$E: = \text{span ext } B \subset C_1: = \{\sum_{i=1}^{m} f_i \mid f_i \in C\} \quad .$$

But every $f = \sum_{i=1}^{m} f_i \in C_1$ is non-injective: for $n > 2^m$

$$\rho_n(f)(\cdot) = \sum_{i=1}^{m} \alpha_i \; \epsilon_i(\cdot)$$

where $\alpha_i \geq 0$ and $\epsilon_i: X_n \to \{-1, +1\}$. Since $(\epsilon_i(k))_{i=1}^{m}$, $k \in X_n$, have only ± 1 entries, there are at most 2^m such different vectors: $n > 2^m$ implies therefore the existence of $k, \ell \in X_n$, $k \neq \ell$, with

$$\epsilon_i(k) = \epsilon_i(\ell) \qquad i = 1,\dots,m$$

and hence

$$\rho_n(f)(k) = \rho_n(f)(\ell) \quad .$$

There are injective functions in F , so $E \subset C_1 \neq F$ - and the example is constructed.

EXERCISES

For a Banach-space E define

$S(E): = \{\varphi \in E' \mid \varphi$ attains its supremum on the closed unit ball $\}$

5.7. Show that $S(c_o) = \{(\eta_n) \in \ell^1 \mid \eta_n = 0$ eventually.$\}$

5.8. Use the example in 5.3. to show, that $S(E)$ need not be a linear space.

5.9. For a σ-finite measure-space show that

$$S(L^1) = \{f \in L^\infty \mid \mu\{x \mid |f(x)| = \|f\|_\infty\} > 0\} .$$

5.10. Find an example of a $\varphi \in (C[0,2])'$ which does not attain its supremum (Hint: $(\chi_{[0,1]} - \chi_{[1,2]})d\mu)$.

5.11. If $S(E)$ is norm-dense in E' , is E reflexive? (Actually E. Bishop and R. Phelps proved that $S(E)$ is norm-dense in E' for all Banach-spaces.)

5.12. A closed subspace G of a normed space E is proximinal if and only if the quotient map

$$\kappa : E \to E/G$$

maps the closed unit ball onto the closed unit ball. (Hint: Look at

$$\|x_0 - y_0\|_E = \inf_{y \in G} \|x_0 - y\|_E = \|\kappa(x_0)\|_{E/G} .)$$

5.13. Reduce the proof of James' theorem to the case of Banach-spaces. (Hint: Locally convex spaces are subspaces of products of Banach-spaces.)

5.14. Show that in James' theorem the assumption of quasicompleteness can be replaced by the closed convex hull $\overline{\text{co}\, A}$ of A being $\mu(E,E')$-complete.

5.15. Find a convex, bounded, non-closed subset of $\mathbb{R}^2$ on which all linear functionals attain their supremum.

5.16. In a quasicomplete locally convex space E, a bounded weakly closed subset A is weakly compact if and only if all $\varphi(A)$, $\varphi \in E'$, are closed.

5.17. In a Banach space, a bounded weakly closed set A is weakly compact if and only if for every weakly closed B which is disjoint from A

$$d(A,B) = \inf\{\|a - b\| \mid a \in A ,\ b \in B\} > 0$$

(Hint: For the necessity look at $B = [\varphi \geq \sup \varphi(A)]$.)(R.C. James)

5.18. A quasicomplete locally convex space is semi-reflexive if and only if every $\varphi \in E'$ attains its supremum on all bounded weakly closed subsets of E - or: on all bounded (absolutely) convex, closed subsets.

5.19. In a quasicomplete locally convex space a weakly closed subset A is weakly compact if and only if for every sequence (x_n) in A there is an $x \in A$ such that for all $\varphi \in E'$

$$\liminf_n \langle\varphi,x_n\rangle \leq \langle\varphi,x\rangle \leq \limsup_n \langle\varphi,x_n\rangle$$

(Hint: Countable compactness for one implication, for the other one, the Dieudonné-Šmulian-theorem.)

§6. PROOF OF THE THEOREM OF R.C.JAMES

6.1. According to the theorem of Eberlein-Grothendieck, a subset A of a quasicomplete locally convex space E is weakly relatively compact if it has interchangeable double-limits with all equicontinuous subsets U^o of E'; this will be used to prove James' theorem. Since A, considered as a space of functions on U^o is uniformly bounded on U^o, the main work for the proof of James' theorem can be accomplished in $\ell^\infty(X)$, the space of bounded (real-valued) functions on a set X.

$\ell^\infty(X)$ carries two topologies: the topology ω_X of pointwise convergence and the norm-topology

$$\|f\|_\infty := \sup_{x\in X} |f(x)|$$

- the topology of uniform convergence on X. The functional

$$s(f) := \sup_{x\in X} f(x)$$

is sublinear, i.e.

(a) $s(f + g) \le s(f) + s(g)$ all $f, g \in \ell^\infty(X)$

(b) $s(\lambda f) = \lambda s(f)$ all $f \in \ell^\infty(X)$ and $\lambda \ge 0$

and continuous with respect to the norm-topology:

$$s(f) \le s(f - g) + s(g) \le \|f - g\|_\infty + s(g)$$

therefore $|s(f) - s(g)| \le \|f - g\|_\infty$.

6.2. Naturally, properties of s can be used to construct functions on X which do not attain their supremum on X.

A preparatory skeleton of the proof is as follows: Arguing by contradiction, assume there is a uniformly bounded sequence of functions (f_n) and a sequence (x_m) in X, such that

$$\lim_n \lim_m f_n(x_m) > \lim_m \lim_n f_n(x_m)$$

and, furthermore, that (f_n) converges pointwise to f. Then

$$\lim_n \lim_m f_n(x_m) - \lim_m f(x_m) > 0$$

and therefore

$$\exists_{n_o} \exists_{c>o} \forall_{n \geq n_o} \exists_{m_o} \forall_{m \geq m_o} \quad (f_n - f)(x_m) \geq c \quad .$$

In particular this will yield

$$s(g - f) \geq c$$

for all $g \in co \{f_n\}$. Certain geometric properties of sublinear functionals will ascertain, for given $\beta_n > 0$, the existence of

$$g_n \in co \{f_m \mid m \geq n\}$$

with

$$s(\sum_1^N \beta_n(g_n - f)) > \beta_N + s(\sum_1^{N-1} \beta_n(g_n - f)) \quad .$$

This, together with the easy fact: $\lim (g_n - f) = 0$, will enforce that the function

$$\sum_1^\infty \frac{1}{n!} (g_n - f)$$

cannot attain its supremum on X.

6.3. Unfortunately equicontinuous sequences in the dual of a locally convex spaces need not have convergent subsequences (1.2.(6)) - but only a cluster-point This fact complicates the proof considerably. Hence, to prepare the main proof two things have to be done:

(1) To find a device to replace limit-points by cluster-points in the above outline of the proof.

(2) To provide some properties of sublinear functionals.

6.4. The first will be essentially solved by the

PROPOSITION (J.D.Pryce): Let (f_n) be a uniformly bounded sequence in $\ell^\infty(X)$ and $D \subset \ell^\infty(X)$ a norm-separable subset. Then there is a subsequence (f_{n_k}) of (f_n) such that the functions $\overline{f}, \underline{f} \in \ell^\infty(X)$

$$\overline{f}(x) := \limsup_k f_{n_k}(x)$$

$$\underline{f}(x) := \liminf_k f_{n_k}(x)$$

satisfy

$$s(h - \underline{f}) = s(h - \overline{f})$$

for all $h \in D$.

Proof: Since s is norm-continuous, D can be assumed to be countable: $D = \{\hat{h}_1, \hat{h}_2, \ldots\}$. Writing this sequence in the form

$$\hat{h}_1, \hat{h}_1, \hat{h}_2, \hat{h}_1, \hat{h}_2, \hat{h}_3, \hat{h}_1, \ldots$$

the resulting sequence (h_n) has the property, that each function h_n returns infinitely often in the sequence.

Since (f_n) is uniformly bounded $\liminf f_n \in \ell^\infty(X)$ exists. Therefore there is an $x_1 \in X$

$$(h_1 - \liminf_m f_m)(x_1) \geq s(h_1 - \liminf_m f_m) - 1$$

and a subsequence $(f_{m,1})$ of (f_m) with

$$\lim_m f_{m,1}(x_1) = \liminf_m f_m(x_1) .$$

Proceeding by induction, there are $x_n \in X$ and a subsequence $(f_{m,n})$ of $(f_{m,n-1})$ such that

$$(h_n - \liminf_m f_{m,n-1})(x_n) \geq s(h_n - \liminf_m f_{m,n-1}) - \frac{1}{n}$$

and

$$\lim_m f_{m,n}(x_n) = \liminf_m f_{m,n-1}(x_n) \quad .$$

The diagonal sequence $(f_{k,k})$ now has the property

$$h_n(x_n) - \limsup_k f_{k,k}(x_n) = h_n(x_n) - \lim_m f_{m,n}(x_n) \geq s(h_n - \liminf_m f_{m,n-1}) - \frac{1}{n} \geq$$

$$\geq s(h_n - \liminf_k f_{k,k}) - \frac{1}{n}$$

With

$$\overline{f} := \limsup_k f_{k,k}$$

$$\underline{f} := \liminf_k f_{k,k}$$

this means

$$s(h_n - \underline{f}) \geq s(h_n - \overline{f}) \geq (h_n - \overline{f})(x_n) \geq s(h_n - \underline{f}) - \frac{1}{n} \quad .$$

But this is, according to the special nature of (h_n), just

$$s(h - \underline{f}) = s(h - \overline{f})$$

for all $h \in \{h_n\} = D$. ∎

This proposition is interesting independently from James-theorem (see Ex. 6.14.).

6.5. Now a property of sublinear functionals: Let E be a vector space and $p: E \to \mathbb{R}$ sublinear

(1) If $\alpha > 0$, $B \subset E$ convex and $v_o \in E$ such that

$$\inf_{y \in B} p(v_o + y) > 1 + p(v_o)$$

then there is a $y_o \in B$ with

$$\inf_{y_1,y_2\in B} p(v_o+y_1+\alpha y_2) > \alpha + p(v_o+y_o) \quad .$$

Proof: Define $1+\delta := \inf p(v_o+y) - p(v_o)$. For $y_1,y_2 \in B$ and

$$z := \frac{y_1+\alpha y_2}{1+\alpha} \in B$$

the following holds

$$(1+\alpha)(v_o+z) = v_o+y_1+\alpha y_2+\alpha v_o$$

$$(1+\alpha)p(v_o+z) \le p(v_o+y_1+\alpha y_2)+\alpha p(v_o)$$

and therefore

$$p(v_o+y_1+\alpha y_2) \ge (1+\alpha)p(v_o+z) - \alpha p(v_o) \ge \alpha+\alpha\delta + \inf p(v_o+y) \ge$$

$$\ge \alpha + \frac{\alpha}{2}\delta + p(v_o+y_o)$$

for an appropriate $y_o \in B$ independent from $y_1,y_2 \in B$. ■

Replacing α by β^{-1} and fixing y_o also on the left side yields

(2) If $\beta > 0$, $B \subset E$ convex and $v_o \in E$ with

$$\inf_{y\in B} p(v_o+y) > 1 + p(v_o)$$

then there is a $y_o \in B$ with

$$\inf_{y\in B} p(\beta(v_o+y_o)+y) > 1 + p(\beta(v_o+y_o)) \quad .$$

This can be easily iterated:

(3) If $\beta_n > 0$, $B_1 \supset B_2 \supset \ldots$ convex subsets of E and

$$\inf_{y \in B_1} p(y) > 1$$

then there are $y_n \in B_n$ with

$$p(\sum_{n=1}^{N} \beta_n y_n) > \beta_N + p(\sum_{n=1}^{N-1} \beta_n y_n) \quad .$$

Proof: It is certainly enough to find $y_n \in B_n$ with

$$\inf_{y \in B_N} p(\frac{1}{\beta_N} \sum_{n=1}^{N-1} \beta_n y_n + y) > 1 + p(\frac{1}{\beta_N} \sum_{n=1}^{N-1} \beta_n y_n) \ .$$

This will be done by induction: For $N = 1$ this is just the assumption. So assume $y_1, \ldots, y_{N-1}$ have been already constructed; apply now (2) with

$$B: = B_N \ , \quad v_o: = \frac{1}{\beta_N} \sum_{n=1}^{N-1} \beta_n y_n \ , \quad \beta: = \frac{\beta_N}{\beta_{N+1}}$$

then

$$\inf_{y \in B_{N+1}} \ldots \geq \inf_{y \in B_N} p(\frac{1}{\beta_{N+1}} \sum_{n=1}^{N} \beta_n y_n + y) > 1 + p(\frac{1}{\beta_{N+1}} \sum_{n=1}^{N} \beta_n y_n) \ . \qquad \blacksquare$$

6.6. The central result, which easily implies James-theorem, reads as follows

THEOREM: Let (L, τ) be a locally convex space, X a set, $T: L \to \ell^\infty(X)$ a linear, τ - ω_X - continuous operator, and $Q \subset L$ a τ - relatively countably compact, convex subset. If all functions Ty, $y \in L$ attain their supremum on X, then TQ is uniformly bounded on X, and TQ and X have interchangeable double-limits

$$TQ \sim X \quad .$$

This theorem, it might be called James' double-limit-theorem, will apply to various other questions. It is important to note that there are three essential assumptions in the theorem

(a) Q is relatively countably compact in L.

(b) Q is convex.

(c) All elements of the linear space L attain their supremum on X.

None of these conditions is dispensable (see Ex. 6.15. - 6.17.) - (c) can be weakened just to make the proof work: see 6.7. Note that the closed convex hull of a compact set need not be compact (Ex. 6.19.) but under certain circumstances which will be investigated in the next paragraph (Krein's theorem).

Proof:

(1) Without loss of generality it can be assumed

$$L \subset \ell^{\infty}(X)$$

$$\tau = \omega_X \Big|_L$$

(2) By Lemma 1.6., Q is contained in a Banach-disc in L, therefore in $\ell^{\infty}(X)$ and is absorbed by all barrels, in particular by the unit ball

$$\{f: X \to \mathbb{R} \mid \|f\|_{\infty} \leq 1\} .$$

This means, Q is uniformly bounded.

(3) For a proof by contradiction against the supremum property of L, it is enough to assume that $f_n \in Q$ and $x_m \in X$ exist with

$$\lim_n \lim_m f_n(x_m) > \lim_m \lim_n f_n(x_m) ,$$

since $-Q$ satisfies also the assumption of the theorem. But the same is true for λQ, so assume that

$$\lim_m [f_n(x_m) - \lim_i f_i(x_m)] \geq 3$$

for all $n \geq n_0$. Taking only a tail of (f_n) yields:

(*) $$\forall_n \ \exists_{m_o} \ \forall_{m \geq m_o} \ f_n(x_m) - \lim_i f_i(x_m) \geq 2 .$$

The proposition 6.4. applied to $D: = \text{span}\{f_n\}$ gives a subsequence (which is not denoted differently) such that

$$\overline{f}(x): = \limsup_i f_i(x)$$

$$\underline{f}(x): = \liminf_i f_i(x)$$

satisfy

$$s(h - \underline{f}) = s(h - \overline{f})$$

for all $h \in D$.

Define $K_n: = \text{co}\{f_m \mid m \geq n\} \subset D$, then

$$s(h - \overline{f}) > 2$$

for all $h \in K_1$, since

$$s(h - \overline{f}) = s(\sum_1^N \lambda_n (f_n - \overline{f})) \geq \sum_1^N \lambda_n (f_n(x_m) - \lim_i f_i(x_m)) \geq$$

$$\geq \sum_1^N \lambda_n 2 = 2$$

by (*) for m sufficiently large.

(4) This is just the situation of the geometric property 6.5.(3) of the sublinear functional s and $B_n: = K_n - \overline{f}$: There are $g_n \in K_n$ with

$$s(\sum_{n=1}^N \beta_n (g_n - \overline{f})) > \beta_N + s(\sum_{n=1}^{N-1} \beta_n (g_n - \overline{f}))$$

(5) $\overline{f}$ and $\underline{f}$ might not be in L: But $K_1 \subset Q$, so the sequence (g_n) in Q has a cluster-point g in L. For each $x \in X$ the sequence $(g_n(x))$ has the cluster-point $g(x)$, therefore

(**) $$-\infty < \liminf_n g_n(x) \leq g(x) \leq \limsup_n g_n(x) < +\infty .$$

Since

$$g_n(x) = \sum_{k=n}^{N} \lambda_k f_k(x)$$

is a convex combination, the following holds:

$$\forall_n \exists_{m_1} \geq n \quad g_n(x) \geq f_{m_1}(x)$$

$$\forall_n \exists_{m_2} \geq n \quad g_n(x) \leq f_{m_2}(x)$$

and hence

$$\underline{f}(x) \leq \liminf_n g_n(x) \leq g(x) \leq \limsup_n g_n(x) \leq \overline{f}(x) .$$

Therefore

$$s(h - \overline{f}) \leq s(h - g) \leq s(h - \underline{f}) = s(h - \overline{f})$$

for all $h \in D$ and in (4) $\overline{f}$ can be replaced by g :

$$s\Big(\sum_{n=1}^{N} \beta_n (g_n - g)\Big) > \beta_N + s\Big(\sum_{n=1}^{N-1} \beta_n (g_n - g)\Big) .$$

(6) Fix now $\beta_n = \frac{1}{n!}$; then

$$\frac{1}{\beta_n} \sum_{m=n+1}^{\infty} \beta_m = \frac{1}{n+1} \Big(1 + \frac{1}{n+2} + \frac{1}{(n+2)(n+3)} + \ldots\Big) \leq \frac{e}{n+1} \to 0 .$$

(7) Since Q is in a Banach-disc, $g \in L$, the sum

$$h_o := \sum_{n=1}^{\infty} \beta_n (g_n - g)$$

converges in L . The function h_o attains therefore its supremum, which will turn out to be impossible:

For this define $h_n := g_n - g$

$$M := \sup_n \sup_{x \in X} |h_n(x)| < \infty$$

and recall that by (5) (**)

$$\liminf_n h_n \leq 0 .$$

Assume now $s(h_o) = h_o(x_o)$, then

$$h_o(x_o) = \sum_1^n \beta_m h_m(x_o) + \sum_{n+1}^{\infty} \beta_m h_m(x_o) \le \underset{\sim\sim\sim\sim\sim\sim\sim\sim\sim\sim\sim\sim\sim\sim\sim}{\sum_1^n \beta_m h_m(x_o) + M \sum_{n+1}^{\infty} \beta_m} .$$

On the other hand

$$h_o(x_o) = s(h_o) \ge s(\sum_1^n \beta_m h_m) - s(-\sum_{n+1}^{\infty} \beta_m h_m) \ge$$

$$\ge \beta_n + s(\sum_1^{n-1} \beta_m h_m) - M \sum_{n+1}^{\infty} \beta_m \ge$$

$$\ge \underset{\sim\sim\sim\sim\sim\sim\sim\sim\sim\sim\sim\sim\sim\sim\sim\sim\sim\sim\sim\sim}{\beta_n + \sum_1^{n-1} \beta_m h_m(x_o) - M \sum_{n+1}^{\infty} \beta_m} .$$

These inequalities yield

$$\beta_n h_n(x_o) \ge \beta_n - 2M \sum_{n+1}^{\infty} \beta_m$$

and therefore

$$h_n(x_o) \ge 1 - 2M \frac{1}{\beta_n} \sum_{n+1}^{\infty} \beta_m \to 1$$

$$\liminf_n h_n(x_o) \ge 1$$

which is a contradiction. ■

6.7. The proof actually dealt with a uniformly bounded sequence in $\ell^{\infty}(X)$. Precisely the following was demonstrated:

Let (f_n) be a uniformly bounded sequence in $\ell^{\infty}(X)$ with the property that for all sequences (g_n), $g_n \in \mathrm{co}\{f_m | m \ge n\}$, there is an ω_X-cluster-point g of (g_n) with

$$\sum_{n=1}^{\infty} \beta_n (g_n - g)$$

exists for all $\beta_n \ge 0$, $(\beta_n) \in \ell^1$, and attains its supremum on X. Then

$$\lim_n \lim_m f_n(x_m) \le \lim_m \lim_n f_n(x_m)$$

whenever all the limits exist. Moreover, if $\sum \beta_n (g_n - g)$ attains also its infimum, then

$$\{f_n\} \sim X \quad .$$

(The latter, since the assumption is obviously true for all subsequences.)

This result is closely related to lemmata obtained by M. De Wilde and S. Simons for their investigations of weakly compact sets. The double-limit-inequality has been used by M.De Wilde to derive some minimax-and maximinimax theorems of S. Simons.

6.8. The James - double-limit-theorem implies the James-theorem:

COROLLARY 1: For a weakly closed subset A of a locally convex space E with the property that $\overline{\mathrm{co}\, A}$ is $\mu(E,E')$-complete the following statements are equivalent:

(1) A is weakly compact.

(2) Every $\varphi \in E'$ attains its supremum on A .

If A is not weakly closed, (2) implies that it is weakly relatively compact.

Proof: By the Eberlein-Grothendieck-theorem it is enough to show that (2) implies that $A \sim U^o$ for all $\mu(E,E')$-neighbourhoods U of zero: the restriction map

$$T:\ L := (E',\ \sigma(E',E)) \rightarrow \ell^{\infty}(A)$$

is well-defined, certainly $\sigma(E',E)$-ω_A-continuous and U^o is $\sigma(E',E)$-compact. James double-limit-theorem gives

$$TU^o \sim A \quad . \qquad \blacksquare$$

In 1972 R.C.James gave a proof of his reflexivity-theorem - "elementary" in the sense, that it is direct and uses neither the double-limit-criterion nor the fact, that a Banach-space is reflexive if and only if its closed unit ball is weakly compact.

6.9. The following lemma collects what is already demonstrated on sets with interchangeable double-limits

LEMMA: Let $Q \subset \ell^\infty(X)$ be uniformly bounded and $Q \sim X$. Then $\overline{Q}^{\omega_X}$ is compact and angelic in the weak topology σ of the normed space $\ell^\infty(X)$ and $\sigma = \omega_X$ on $\overline{Q}^{\omega_X}$.

Proof: If X is equipped with the discrete topology then $\ell^\infty(X) = C^{\ell}(X) = C(\beta X)$ as normed spaces: the result in 4.4. states that Q is weakly relatively compact therefore $\overline{Q}^{\sigma}$ is compact and $= \overline{Q}^{\omega_X}$. The weak topology of a normed space is angelic by 3.10.(1). ∎

6.10. Together with the main theorem this yields

COROLLARY 2: Let L be a subspace of $\ell^\infty(X)$ such that all $f \in L$ attain their supremum on X. If $Q \subset L$ is convex and ω_X-relatively countably compact in L, then

(1) $\overline{Q}^{\omega_X}$ is uniformly bounded on X, compact and angelic in the weak topology σ of the normed space $\ell^\infty(X)$.

(2) $\sigma = \omega_X$ on $\overline{Q}^{\omega_X}$.

(3) In particular, Q is ω_X-relatively sequentially compact in L.

That Q is ω_X-relatively compact (in $\ell^\infty(X)$) is not astonishing (see Ex. 6.13.). If X is a topological space, then $C^{\ell}(X)$ is a norm-closed $= \sigma$-closed subspace of $\ell^\infty(X)$, therefore if $L \subset C^{\ell}(X)$

$$\overline{Q}^{\omega_X} \subset C^{\ell}(X)$$

- so this is another approach to find results as those in 4.9..

6.11. What is happening if the functions attain their supremum on X always on a subset $Y \subset X$?

COROLLARY 3: If $Y \subset X$, L a subspace of $\ell^\infty(X)$, such that all $f \in L$ attain their supremum on Y, then every convex, ω_Y-relatively countably compact subset Q of L is ω_X-relatively countably compact in L. On $\overline{Q}^{\omega_X}$ the topologies ω_X, ω_Y, $\sigma(\ell^\infty(Y), \ell^\infty(Y)')$, and $\sigma(\ell^\infty(X), \ell^\infty(X)')$ coincide.

This is a sort of a maximum principle for L.

In particular, $\overline{Q}^{\omega_X}$ is angelic in these topologies and statements (1) - (3) of Corollary 2 above hold.

Proof: First note, that the restriction map

$$L \to \ell^\infty(Y)$$

is one-to-one by the assumption (so ω_Y is a Hausdorff-topology on L) and Corollary 2 applies: Q is therefore $\sigma(\ell^\infty(Y), \ell^\infty(Y)')$-relatively countably compact in L. The embedding

$$L \to \ell^\infty(X)$$

is, again by the assumption, an isometry if L is equipped with $\| \|_Y$ and $\ell^\infty(X)$ with $\| \|_X$ - therefore continuous with respect to the associated weak topologies, hence Q is $\sigma(\ell^\infty(X), \ell^\infty(X)')$-relatively countably compact in L, in particular ω_X-relatively countably compact in L.

Furthermore, by Corollary 2, $\omega_X = \sigma(\ell^\infty(X), \ell^\infty(X)')$ on the $\sigma(\ell^\infty(X), \ell^\infty(X)')$-compact set

$$\overline{Q}^{\omega_X}$$

therefore the two coarser topologies ω_Y and $\sigma(\ell^\infty(Y), \ell^\infty(Y)')$ also coincide. ∎

A special consequence is (S. Simons), that a sequence in Q converges pointwise on Y if and only if it converges pointwise on X (see also Ex. 6.20, Ex. 8.19.) Certainly the assumptions on Q (and L) may be replaced by conditions which imply 6.7. .

EXERCISES

6.12. If $f_n: X \to \mathbb{R}$ is a sequence which converges pointwise to zero then every sequence (g_n)

$$g_n \in \text{co}\{f_m \mid m \geq n\}$$

converges pointwise to zero.

6.13. (a) Every uniformly bounded subset of $\ell^\infty(X)$ is ω_X-relatively compact.

(b) When are ω_X and $\sigma(\ell^\infty(X), \ell^1(X))$ angelic? (Hint: 3.22.)

(c) $A \subset \ell^\infty(X)$ is $\sigma(\ell^\infty,(\ell^\infty)')$-relatively compact if and only if it is uniformly bounded and $A \sim X$. (Hint: 6.9. and the Eberlein-Grothendieck-theorem.)

(d) Deduce from (c): If E is a norm-closed subspace of $\ell^\infty(X)$ then $A \subset E$ is $\sigma(E,E')$-relatively compact if and only if it is bounded (in the Banach space E) and $A \sim X$. (This result is due to S. Simons.)

6.14. If $G \subset \ell^\infty(X)$ is a norm-separable subset and, $B \subset G$ uniformly bounded, then B is ω_X-relatively countably compact in G if and only if it is ω_X-relatively sequentially compact in G. (Hint: Apply 6.4 to $D: = \overline{\{f_n\}}$ and show that the subsequence actually converges to a cluster-point in D.)

6.15. Show that a convex uniformly bounded subset of $L \subset \ell^\infty(X)$, L a vector space of functions which attain their supremum on X, need not have interchangeable double-limits with X. (Hint: Take the unit ball in $C(X)$, X compact.)

6.16. Give an example of an ω_X-convergent uniformly bounded sequence (f_n), situated in a subspace $L \subset \ell^\infty(X)$ of functions, which attain their supremum on X, but $\{f_n\}$ and X do not have interchangeable double-limits. (Hint: Change the example 4.13. in such a way that $f_n(x) = 1$ for $x \geq 1 - (2n)^{-1}$ and set $L: = \text{span } f_n$; observe 4.4. or check directly.)

6.17. There is a convex ω_X-compact subset $Q \subset \ell^\infty(X)$, $X = [-1, \infty[$, such that all $f \in Q$ attain their supremum and infimum, but Q and X do not have interchangeable double-limits. (Hint: Take the functions f_c from 1.18., change them in $[-1,0]$ to $f_c(x): = -x$, look at the ω_X-closed convex hull of $\{f_c \mid o \leq c \leq 1\}$).

6.18. (a) What are the simplifications in the proof of the James-double-limit-theorem, if Q is τ-relatively sequentially compact? For which spaces E can the James-theorem be proven with this simpler result?

(b) (W. Govaerts) If E is a separable Banach-space, $A \subset E$ bounded and not weakly compact, then there is a sequence (φ_n) in E' such that

$$\varphi_n \to 0 \big|_{\sigma(E',E)}$$

$$\sup_{x \in A} \langle \varphi_n, x \rangle \geq 2$$

(Hint: Apply Grothendieck's completeness criterion to $z \in \bar{A}^{\sigma(E'',E')} \setminus E$.) This, together with (4), (6), and (7) of the proof of the main theorem 6.6., gives directly James' theorem for separable Banach spaces.

6.19. Give an example of a zero-sequence (x_n) in a locally convex space and $(\xi_n) \in \ell^1$ such that

$$\sum_{n=1}^{\infty} \xi_n x_n$$

does not converge. Deduce from this, that the closed convex hull of a compact set need not be compact. (Hint: Take the space of all finite-sequences with the ℓ^1-norm and consider $n^{-1} e_n$, e_n the n-th unit vector.)

6.20. (S.Simons) If X is pseudo-compact, then a uniformly bounded sequence (f_n) converges pointwise if and only if it converges weakly in the normed space $(C(X), \|\cdot\|_\infty)$. (Hint: 6.7., 6.12., 2.13., and 6.9.)

6.21. Check the content of Corollary 3 (6.11.) in the case where X is the closed unit disc in $\mathbb{R}^2$, Y its boundary and L the space of harmonic (in X^0) functions which are continuous on X.

§7. APPLICATIONS OF THE SUP-THEOREM

7.1. It was disturbing already several times: The closed convex hull of a compact set need not be compact (6.19.). The main reason for this is, that for $\xi_n \geq 0$, $\Sigma\xi_n = 1$, $x_n \in A$ (A compact, $0 \in A$) the series

$$\sum_{n=1}^{\infty} \xi_n x_n$$

is certainly Cauchy, its partial sums are in $\operatorname{co} A$, but the space may not be "complete" enough to contain the limit.

THEOREM (M. Krein): Let E be a locally convex space and $A \subset E$ $\sigma(E,E')$-relatively compact. Then the closed convex hull $\overline{\operatorname{co} A}$ is $\sigma(E,E')$-compact if and only if it is $\mu(E,E')$-complete.

Proof: Every weakly compact set is weakly complete and therefore complete for all locally convex topologies which are compatible with the dual system, in particular for the Mackey-topology $\mu(E,E')$. Conversely, assume that $\overline{\operatorname{co} A}$ is $\mu(E,E')$-complete. Since

$$\sup_{x \in \overline{\operatorname{co} A}} \langle \varphi, x \rangle = \sup_{x \in \overline{A}} \langle \varphi, x \rangle$$

for every $\varphi \in E'$ and $\overline{A} \subset \overline{\operatorname{co} A}$ is weakly compact, each $\varphi \in E'$ attains its supremum on $\overline{\operatorname{co} A}$: James' Theorem 6.8. implies that $\overline{\operatorname{co} A}$ is $\sigma(E,E')$-compact. ■

In particular: In quasicomplete spaces the closed convex hull and the closed absolute convex hull $\overline{\Gamma A} = \overline{\operatorname{co}(A \cup (-A))}$ of a weakly relatively compact set A are weakly compact.

7.2. A similar result holds for other topologies than the weak one.

COROLLARY: Let (E,τ) be a locally convex space and $A \subset E$ τ-relatively compact: $\overline{\operatorname{co} A}$ is τ-compact if and only if it is $\mu(E,E')$-complete.

Proof: If $\overline{\operatorname{co} A}$ is $\mu(E,E')$-complete, $\overline{\operatorname{co} A}$ is $\sigma(E,E')$-compact by the theorem and therefore τ-complete. But $\overline{\operatorname{co} A}$ is τ-precompact (ex. 1.13.) and hence τ-compact. ■

Certainly, for a τ-relatively compact set A, the closed convex hull $\overline{\text{co}\,A}$ is τ-compact if it is τ-complete, but the assumption of $\mu(E,E')$-completeness is a priori weaker.

7.3. If A and B are closed convex subsets of a locally convex space, their sum

$$A + B := \{a + b \mid a \in A, b \in B\}$$

is convex, but need not be closed (= weakly closed), see Ex. 7.14., but it is, if A is weakly compact. Therefore, in a semi-reflexive space the sum of two closed, bounded, convex sets is always closed. Is this a characterization of semi-reflexive spaces? It will be shown, that it is. How obscure are to be the sets A and B, such that $A+B$ is not closed? In Banach-spaces, it can happen, that the sum of the closed unit ball with the closed unit ball of an equivalent norm is open (M. Edelstein - A. C. Thompson, see 7.18.)

Moreover, in c_0 there is a closed hyperplane H (through 0) such that $B + B \cap H$, B the closed unit ball, is not closed (Ex. 7.16.). An analysis of this example yields the following:

LEMMA: Let $I = [-1, +1]$ the unit interval, F a locally convex space, $A \subset F$ absolutely convex, $\varphi \in F'$ such that

$$\sup_{x \in A} \langle \varphi, x \rangle = 1$$

and this supremum is not attained. Then

$$A \times I + [(A \times I) \cap \text{graph}\,\varphi] \subset F \times \mathbb{R}$$

is not closed (in $F \times \mathbb{R}$ with the product topology.)

Proof: $\text{graph}\,\varphi := \{(x,\alpha) \in F \times \mathbb{R} \mid \langle \varphi, x \rangle = \alpha\}$. Therefore by the assumption on the supremum, $(x,\alpha) \in (A \times I) \cap \text{graph}\,\varphi$ implies $\alpha < 1$. In particular

$$(0,2) \notin A \times I + [(A \times I) \cap \text{graph}\,\varphi] .$$

On the other hand, since A is absolutely convex, for every $0 < \varepsilon < 1$ there are $x_\varepsilon \in A$ with

$$\langle \varphi, x_\varepsilon \rangle = 1 - \varepsilon$$

such that

$$(0,2) = \lim_{\varepsilon\to 0} [(-x_\varepsilon,1) + (x_\varepsilon,\langle\varphi,x_\varepsilon\rangle)] \in \overline{A \times I + [(A \times I) \cap \text{graph } \varphi]}.$$ ■

This gives a characterization of weakly compact sets in terms of sums:

THEOREM: In a locally convex space F an absolutely convex, bounded, $\mu(E,E')$-complete subset A is weakly compact if and only if for every $\varphi \in F'$ the set

$$A \times I + [(A \times I) \cap \text{graph } \varphi]$$

is closed in $F \times \mathbb{R}$. $(I = [-1, +1])$.

Proof: Since $A \times I$ is weakly compact when A is, and graph φ is closed, the condition is necessary. Conversely, by James' theorem 6.8. if there were a $\varphi_0 \in F'$ which does not attain its supremum on A (which is 1 without loss of generality), the lemma would imply that

$$A \times I + [(A \times I) \cap \text{graph } \varphi].$$

were not closed. ■

7.4. This gives another characterization of semi-reflexive spaces

COROLLARY 1: A quasicomplete locally convex space E is semi-reflexive if and only if for every bounded, absolutely convex, closed subset $B \subset E$ and every closed zero-hyperplane H the set

$$B + (B \cap H)$$

is closed.

Proof: As was mentioned before it is enough to show, that this is sufficient. Decompose $E = F \times \mathbb{R}$ topologically then F is quasicomplete and it is enough to show that F is semi-reflexive, i.e. that every absolutely convex, bounded, closed set $A \subset F$ is weakly compact: But this is immediate by the theorem, since graphφ is a hyperplane in E for every $\varphi \in F'$ and $B = A \times I$ is absolutely convex, closed and bounded. ■

The following result is now immediate

COROLLARY 2 (V.Klee): A quasicomplete locally convex space is semi-reflexive if and only if $A + B$ is closed for all bounded, closed, convex sets A and B.

7.5. For Banach-spaces E these results can be sharpened: If B is the closed unit ball and F an arbitrary (closed zero-) hyperplane then E is reflexive if and only if $A := B \cap F$ is weakly compact (in F). But $A \times I$ is the closed unit ball of an equivalent norm on E, therefore the theorem gives the

COROLLARY 3: A Banach-space is non-reflexive if and only if there is an equivalent norm (with closed unit ball C) and a closed zero-hyperplane H such that

$$C + (C \cap H)$$

is not closed.

Though the actual change of the norm in this result is little:

$$|||(x,\alpha)||| := \max (\|x\|_E , |\alpha|)$$

for $(x,\alpha) \in F \times \mathbb{R} = E$ (and often there are hyperplanes in E such that $||| \; ||| = \| \; \|$, see Ex. 7.17.; but not always: strictly convex spaces), it is not known whether Corollary 3 holds without passing to an equivalent norm.

7.6. The sum and the convex hull of two sets are related, so it might not be astonishing that rather the same results hold for the convex hull: First for two convex sets A and B in a locally convex space the convex hull

$$co(A \cup B) = \{\lambda a + (1-\lambda)b \mid a \in A,\ b \in B,\ o \leq \lambda \leq 1\}$$

is closed if A is compact and B closed (Ex. 7.20.)

LEMMA: Let A be an absolutely convex, closed, bounded subset of a locally convex space F, $\varphi \in F'$ such that the supremum

$$\sup_{x \in A} \langle \varphi, x \rangle = 1$$

is not attained, then

$$\mathrm{co}[(A \times \{0\}) \cup ((A \times I) \cap \mathrm{graph}\ \varphi)]$$

is not closed in $F \times \mathbb{R}$. $(I = [-1, +1])$.

Proof: The element $(0, \frac{1}{2})$ will be in the closure but not in the hull: Assume

$$(0,\tfrac{1}{2}) = \lambda_1(x,0) + \lambda_2(y,\alpha) \in \mathrm{co}\ \ldots$$

$(x,y \in A,\ \langle\varphi,y\rangle = \alpha < 1,\ \lambda_1 + \lambda_2 = 1,\ \lambda_i \geq 0)$, then $\lambda_2\,\alpha = \frac{1}{2}$ and $\lambda_1 x + \lambda_2 y = 0$, therefore

$$0 = \lambda_1\langle\varphi,x\rangle + \lambda_2\langle\varphi,y\rangle = \lambda_1\langle\varphi,x\rangle + \tfrac{1}{2} \ .$$

Since $|\langle\varphi,x\rangle| < 1$ it follows that $\lambda_1 > \frac{1}{2}$ and therefore $\lambda_2 < \frac{1}{2}$: but this contradicts $\lambda_2\alpha = \frac{1}{2}$.

Furthermore, as in 7.3., there are $x_\varepsilon \in A$ with $\langle\varphi,x_\varepsilon\rangle = 1-\varepsilon$ and hence

$$(0,\tfrac{1}{2}) = \lim_{\varepsilon\to 0}\ (\tfrac{1}{2}(-x_\varepsilon,0) + \tfrac{1}{2}(x_\varepsilon,\ 1-\varepsilon)) \in \overline{\mathrm{co}\ \ldots} \qquad \blacksquare$$

More or less in the same way as in 7.3. and 7.4. this lemma yields the

THEOREM: In a locally convex space F an absolutely convex bounded $\mu(E,E')$-complete set A is weakly compact, if and only if for every $\varphi \in F'$ the convex hull of $A \times \{0\}$ and $(A \times I) \cap \mathrm{graph}\ \varphi$ is closed in $F \times \mathbb{R}$. $(I = [-1, +1])$.

COROLLARY 1: A quasicomplete locally convex space is semi-reflexive if and only if there exists a closed zero-hyperplane H_1, such that for all closed zero-hyperplanes H_2 and all absolutely convex, bounded closed sets B the convex hull

$$\mathrm{co}[(B \cap H_1) \cup (B \cap H_2)]$$

is closed.

In particular

COROLLARY 2 (V.Klee): A quasicomplete locally convex space is semi-reflexive if and only if the convex hull of each two closed, convex, bounded sets is closed.

For Banach-spaces see Ex. 7.20. - 7.22.

7.7. The Hahn-Banach-separation-theorem states that a closed convex A and a compact convex B (disjoint from each other) are in different open half-spaces:

$$A \subset [\varphi < \alpha] \qquad B \subset [\varphi > \alpha]$$

That this is always possible, is characteristic for semi-reflexive spaces:

THEOREM: A quasicomplete locally convex space E is semi-reflexive if and only if each two disjoint closed, bounded, convex sets can be strictly separated.

Proof: It is sufficient to show, that every absolutely convex bounded closed subset B is weakly compact, that is, by James' theorem, that every $\varphi_o \in E'$ attains its supremum on B. Assume the contrary, then there is a $\varphi_o \in E'$

$$\sup_{x \in B} \langle \varphi_o , x \rangle = 1$$

and this is not attained. By assumption, there is a $\varphi_1 \in E'$ with

$$B \subset [\varphi_1 < 1] \quad \text{and} \quad A := [\varphi_o = 1] \cap 4B \subset [\varphi_1 > 1]$$

(A is non-empty!) If it can be shown that $\varphi_o = \varphi_1$ on $\operatorname{span} B =: F$, a contradiction is produced.

(a) F equipped with the gauge-norm of B is a normed space and B is its closed unit ball: $F \hookrightarrow E$ is continuous therefore φ_o and φ_1 are both continuous on F. Since

$$\sup_{x \in B} \langle \varphi_o , x \rangle = 1$$

$\|\varphi_o\| = 1$ and, because of $B \subset [\varphi_1 < 1]$, $\|\varphi_1\| \leq 1$.

For $\varepsilon \in]0,\frac{1}{2}]$ take $x_\varepsilon \in F$, $\|x_\varepsilon\| = 1$, and $\langle \varphi_o , x_\varepsilon \rangle = 1 - \varepsilon$. Then $(1-\varepsilon)^{-1} x_\varepsilon \in A$ and hence

$$\langle \varphi_1 , (1-\varepsilon)^{-1} x_\varepsilon \rangle > 1$$

$$\langle \varphi_1 , x_\varepsilon \rangle > 1 - \varepsilon$$

and $\|\varphi_1\| = 1$ is proven.

(b) Assume that the linear functionals φ_o and φ_1 are not multiples of each other (on F), then their kernels are different, i.e. there is an $x_o \in F$ with

$$\langle \varphi_o, x_o \rangle = o \quad \text{and} \quad \langle \varphi_1, x_o \rangle = 1 \quad .$$

Take the x_ε from (a) and $\eta = 2\|x_o\|^{-1}$ then

$$z_\varepsilon := (1-\varepsilon)^{-1} x_\varepsilon - \eta x_o \in A$$

and therefore

$$1 < \langle \varphi_1, z_\varepsilon \rangle = \langle \varphi_1, (1-\varepsilon)^{-1} x_\varepsilon \rangle - \eta \le \|\varphi_1\| (1-\varepsilon)^{-1} \|x_\varepsilon\| - \eta =$$
$$= (1-\varepsilon)^{-1} - \eta$$

which is certainly not true for small ε.

(c) This means $\varphi_o = \alpha \varphi_1$; since $\|\varphi_o\| = \|\varphi_1\|$ it follows $\alpha = \pm 1$ and then, by the separation property, $\alpha = 1$. ∎

There is a stronger theorem by V. Klee stating, that a quasicomplete locally convex space is already semi-reflexive, if each two closed, bounded, convex, disjoint sets A and B can be separated, i.e. there is a $\varphi \in E'$, $\alpha \in \mathbb{R}$

$$A \subset [\varphi \le \alpha] \quad \text{and} \quad B \subset [\varphi \ge \alpha] \quad .$$

The proof of the above theorem actually demonstrated an "individual" result:

THEOREM: An absolutely convex, bounded, closed subset B of a quasicomplete locally convex space is weakly compact if and only if every convex, bounded, closed set A, which is disjoint from B, can be strictly separated from B.

7.8. If E is a locally convex space, X a set, $\mathfrak{A}$ a σ-ring of subsets of X, then an E-valued measure m is a map $\mathfrak{A} \to E$ which is σ-additive, i.e.

$$m\left(\bigcup_{n=1}^{\infty} A_n\right) = \sum_{n=1}^{\infty} m(A_n)$$

for all mutually disjoint sequences (A_n), $A_n \in \mathfrak{A}$.

THEOREM: The range

$$m(\mathfrak{A}) := \{m(A) \mid A \in \mathfrak{A}\}$$

of every E-valued measure m, E a quasicomplete locally convex space, is weakly relatively compact.

This was first proved by R.G. Bartle, N. Dunford, and J. Schwartz for Banach-spaces; the following proof is due to I. Tweddle.

Proof: It is certainly enough to prove the theorem for complete spaces E, since weakly compact sets are bounded.

(a) First, for $X_o \in \mathfrak{A}$, the relative range

$$m(\mathfrak{A}(X_o)) := \{m(A) \mid A \in \mathfrak{A},\ A \subset X_o\}$$

is weakly relatively compact: To show this, according to James' theorem, take a $\varphi \in E'$. Then the finite (real-valued) measure $\varphi \circ m$ has a Hahn-decomposition: $X_o = X_+ \cup X_-$, $X_+ \cap X_- = \emptyset$, $X_+, X_- \in \mathfrak{A}$ and the restrictions

$$\varphi \circ m|_{X_+} \quad \text{and} \quad -\varphi \circ m|_{X_-}$$

are non-negative measures. Hence for every $A \in \mathfrak{A}$, $A \subset X_o$

$$\langle \varphi, m(A) \rangle = \varphi \circ m(A \cap X_+) + \varphi \circ m(A \cap X_-) \leq \varphi \circ m(A \cap X_+) \leq$$

$$\leq \langle \varphi, m(X_+) \rangle$$

and φ attains its supremum on $m(\mathfrak{A}(X_o))$.

(b) The Eberlein-Grothendieck-theorem finishes now the proof: take a sequence $(m(A_n))_n$ in $m(\mathfrak{A})$, then

$$A_n \subset \bigcup_{\ell=1}^{\infty} A_\ell =: X_o$$

therefore $m(A_n) \in m(\mathfrak{A}(X_o))$ - which is weakly relatively compact: $(m(A_n))_n$ has a weak cluster-point. ∎

It is enough to assume that m is a measure with respect to the weak topology; but a theorem of W. Orlicz and B.J. Pettis states that "weak" measures are measures for all topologies compatible with the dual system - sometimes even for other topologies: they all were characterized by P. Dierolf.

It can be shown, that under certain circumstances the range of a vector-measure is even convex and weakly compact. This result, usually called Liapunov-theorem, implies bang-bang-principles in optimal control theory (see I. Kluvanek - G. Knowles).

7.9. The Schauder-fixed-point-theorem states that every continuous mapping

$$\Phi: A \to A$$

A a compact convex subset of a locally convex space, has a fixed point. So it might be possible in convex closed, non-compact sets to construct mappings without fixed points: Again James' theorem will help

PROPOSITION: Let (E,τ) be a locally convex space $A \subset E$ a bounded, convex subset, $0 \in A$ and $\varphi \in E'$ such that $\langle \varphi, x \rangle \geq -1$ for all $x \in A$ and the supremum

$$\sup_{x \in A} \langle \varphi, x \rangle = 1$$

is not attained. Then there is a $\sigma(E,E') - \tau$ continuous function

$$\Phi: A \to A$$

with

$$\langle \varphi, \Phi(x) \rangle = \tfrac{1}{2}(\langle \varphi, x \rangle + 1) \quad .$$

In particular, Φ has no fixed-point in A.

The following construction was carried out by A. Cellina for the unit ball of a non-reflexive Banach-space to disprove the Peano-theorem (see 7.10. below) there.

Proof: Take a sequence $\alpha_n \in]0,1[$, $n = 0,1,2,\ldots$, $\alpha_n \uparrow 1$, continuous functions $p_n: \mathbb{R} \to [0,1]$ with

$$\operatorname{supp} p_1 \subset]-\infty, 2\alpha_1 - 1]$$

$$\operatorname{supp} p_n \subset [2\alpha_{n-2} - 1, 2\alpha_n - 1[\qquad n = 2,3,\ldots$$

$$\sum_{n=1}^{\infty} p_n(t) = 1 \quad \text{for} \quad t \in]-\infty, 1[$$

and $x_n \in A$ with $\langle \varphi, x_n \rangle = \alpha_n$. Define now for $x \in A$

$$\Phi(x) := (\langle \varphi, x \rangle + 1) \sum_{n=1}^{\infty} (2\alpha_n)^{-1} p_n(\langle \varphi, x \rangle) x_n$$

-the series is well-defined, since for every $x \in A$ there is a neighbourhood (in the weak topology) such that the sum has at most three terms; in particular Φ is $\sigma(E,E')$-τ-continuous.

Furthermore if $p_n(\langle \varphi, x \rangle) \neq 0$ then $\langle \varphi, x \rangle \leq 2\alpha_n - 1$ and therefore

$$(\langle \varphi, x \rangle + 1)(2\alpha_n)^{-1} \leq 1 \quad .$$

Since $0 \in A$, this means that $\Phi(x)$ is a finite convex combination of points in A: $\Phi(A) \subset A$. Recalling $\langle \varphi, x_n \rangle = \alpha_n$ yields

$$\langle \varphi, \Phi(x) \rangle = (\langle \varphi, x \rangle + 1) \sum_n (2\alpha_n)^{-1} p_n(\langle \varphi, x \rangle) \langle \varphi, x_n \rangle =$$

$$= (\langle \varphi, x \rangle + 1) \cdot \tfrac{1}{2} \quad .$$

If Φ had a fixed-point: $\Phi(\bar{x}) = \bar{x} \in A$, then

$$\langle \varphi, \bar{x} \rangle = \langle \varphi, \Phi(\bar{x}) \rangle = \tfrac{1}{2} (\langle \varphi, \bar{x} \rangle + 1)$$

therefore $\langle \varphi, \bar{x} \rangle = 1$ contradicting the assumption. ∎

James' theorem , some obvious shifts such that the assumptions of the proposition are satisfied, together with Schauder's fixed-point-theorem give the

THEOREM: In a locally convex space E a bounded convex $\mu(E,E')$-complete subset A is weakly compact if and only if every weakly continuous mapping

$$\Phi : A \to A$$

has a fixed point.

V. Klee proved [33], that in a metrizable locally convex space, a convex set A is compact if and only if every continuous $\Phi: A \to A$ has a fixed point. It seems that it is not known whether this fixed-point characterization of compact (convex, closed) sets is true for other than weak or metrizable topologies.

There is a related result due to D. P. and V. D. Milman (for a proof see R. C. James [30]): In a locally convex space a complete, bounded convex subset A is weakly compact if and only if for every closed convex subset $B \subset A$ each affine continuous map $B \to B$ has a fixed point.

7.10. The Peano-theorem says, that if E is a finite-dimensional Banach-space

$$f: \mathbb{R} \times E \to E$$

continuous, then the initial value problem

$$x'(t) = f(t,x(t))$$
$$x(0) = 0$$

has locally a solution.

THEOREM (A.Cellina): Peano's theorem does not hold for non-reflexive Banach-spaces E .

Actually, Peano's theorem does not hold in any infinite-dimensional Banach-space [19]

Proof: James' theorem ascertains the existence of a $\varphi \in E'$, $\|\varphi\| = 1$, such that φ does not attain its supremum on the closed unit ball B . By the proposition 7.9. there is a (norm-)continuous function

$$\Phi: B \to B$$

with

$$(*) \qquad \langle\varphi, \Phi(x)\rangle = \tfrac{1}{2}(\langle\varphi, x\rangle + 1) \quad .$$

Then

$$\tilde{\Phi}(x) := \begin{cases} \Phi(x) & \|x\| \le 1 \\ \Phi(\frac{x}{\|x\|}) & \|x\| \ge 1 \end{cases}$$

is a continuous extension of Φ to the whole space E and $f: \mathbb{R} \times E \to E$

$$f(t,x):=\begin{cases}0 & t=0\\ 2t\,\tilde{\Phi}(t^{-2}x) & t\neq 0\end{cases}$$

is also continuous since always $\|\Phi(y)\| \leq 1$.

(a) Assume $x: [a,b] \to E$, $0 \in [a,b]$, $a < b$, is a solution of the initial value problem

$$x'(t) = f(t,x(t)) \quad t \in [a,b]$$
$$x(0) = 0 \quad .$$

If $0 < b \leq 1$ (the other case can be treated in the same way) then the following holds

$$x(t) = x(t) - x(0) = \int_0^t x'(s)ds = \int_0^t f(s,x(s))ds$$

$$\|x(t)\| \leq \int_0^t \|f(s,x(s))\|ds \leq \int_0^t 2sds = t^2$$

(Riemann-integrals) therefore $\|x(t)\| \leq 1$ for $t \leq b$, such that (for $t \neq 0$)

$$\frac{d}{dt}\langle\varphi,x(t)\rangle = \langle\varphi,x'(t)\rangle = \langle\varphi,f(t,x(t))\rangle =$$
$$= \langle\varphi,2t\,\Phi(t^{-2}x(t))\rangle = 2t\,\tfrac{1}{2}(\langle\varphi,t^{-2}x(t)\rangle + 1) =$$
$$= \frac{1}{t}\langle\varphi,x(t)\rangle + t$$

because of (*) . Since

$$|\langle\varphi,x(t)\rangle| \leq \|x(t)\| \leq t^2$$

$u(t):=\langle\varphi,x(t)\rangle$ is a solution of the following problem

$$(**)\begin{cases}u'(t) = \dfrac{u(t)}{t} + t & t \in \,]0,b[\\ |u(t)| \leq t^2 & \end{cases}$$

(b) The only solution of (**) is $u(t) = t^2$: For this, consider the non-negative function

$$v(t):= t^2 - u(t) \geq 0 \quad ;$$

it satisfies the differential equation

$$v'(t) = \frac{v(t)}{t}$$

If there were a $t_o \in]0,b]$ with $v(t_o) > 0$, then $v(t) = ct$, $c > 0$, and therefore

$$|u(t)| = |t^2 - ct| \leq t^2 \qquad t \in]0,b]$$

which is impossible.

(c) This means that

$$\langle \varphi, x(t) \rangle = t^2$$

for $t \in]0,b]$, $\|x(t)\| \leq t^2$, and therefore

$$\langle \varphi, t^{-2}x(t) \rangle = 1 \quad \text{and} \quad t^{-2}x(t) \in B$$

- contradicting the assumption on the supremum. ■

EXERCISES

7.11. Let μ be a Radon-measure on a compact set K, E a quasicomplete locally convex space and $f: K \to E$ continuous. Then there is an $x \in E$ with

$$\langle \varphi, x \rangle = \int_k \langle \varphi, f(s) \rangle \mu(ds)$$

for all $\varphi \in E'$. (Hint: Show that $[\varphi \rightsquigarrow \int_K \langle \varphi, f \rangle d\mu] \in E'^*$ is in $f(K)^{oo}$ and use Krein's theorem.)

7.12. A Banach-space is reflexive if and only if it is the union of countably many weakly compact sets. (Hint: Krein's theorem and the Baire category-theorem.)

7.13. If A is a bounded subset of a locally convex space E and $A \sim Q$ for all equicontinuous sets in E', then $\overline{\Gamma A} \sim Q$ for all equicontinuous sets. (Hint: Apply the theorems of Eberlein-Grothendieck and Krein in the completion of E.)

7.14. Find two closed convex sets $A, B \subset \mathbb{R}^2$ such that $A+B$ is not closed.

7.15. The sum of two closed subsets of a locally convex space, one of which is compact, is closed.

7.16. If B is the closed unit ball of c_o and

$$H = \{(\xi_n) \in c_o \mid \sum_{n=1}^{\infty} \xi_n 2^{-n} = 0\}$$

then $B + B \cap H$ is not closed. (Hint: Write $c_o = \mathbb{R} \times F$ by isolating the first component, $H = \text{graph}\,\varphi$, $\varphi \in F'$, and use lemma 7.3.)

7.17. Show that the Banach spaces $E = c_o, \ell^\infty, L^\infty$ (on a measure-space with an atom) admit a hyperplane F such that $E = F \times \mathbb{R}$ holds isometrically if $F \times \mathbb{R}$ has the maximum-norm, see 7.5.

7.18. Let E be a locally convex space

(a) If A is a convex subset of E with non-empty interior A^o, then A is open if and only if for every $0 \neq \varphi \in E'$ the set $\varphi(A)$ is open.

(b) If B is absolutely convex, $B^o \neq \emptyset$, and C convex then $B+C$ is open if and only if for every $0 \neq \varphi \in E'$ either $\varphi(B)$ or $\varphi(C)$ is open.

(c) If B is the closed unit ball of a Banach space E and $T: E \to E$ a bijective continuous operator, then $B + T^{-1}B$ is open if and only if

$$T'(S(E)) \cap S(E) = \{0\}$$

(Hint: 5.7. for the definition of $S(E)$, 5.16., and $\langle \varphi, T^{-1}B \rangle = \langle (T')^{-1}\varphi, B \rangle$.)

These results are due to M. Edelstein and M.C. Thompson, who actually constructed such an operator in c_o, a space where $S(E)$ is linear and has a particularly simple description (5.7.). In terms of approximations (c) reads as follows:

(d) If B is the closed unit ball of a Banach space, C convex and closed, and $x \in \overline{(B+C)} \setminus (B+C)$, then x has no best-approximation in C.

7.19. There are closed subspaces G, H in $C([0,1])$, $G \subset H$, such that $B \cap G + B \cap H$, B the closed unit ball, is not closed. (Hint: Every separable Banach-space is isometrically embedded into $C([0,1])$.)

7.20. If A and B are closed convex subsets of a locally convex space, A compact, then $co(A \cup B)$ is closed.

7.21. Find closed zero-hyperplanes H_1 and H_2 in c_0 such that $co[(B \cap H_1) \cup (B \cap H_2)]$ B the closed unit ball, is not closed. (Hint: 7.16., 7.6.)

7.22. A Banach-space is non-reflexive if and only if there is an equivalent norm with closed unit ball B_1 and closed zero-hyperplanes H_1 and H_2 such that $co[(B_1 \cap H_1) \cup (B_1 \cap H_2)]$ is not closed.

7.23. Construct in c_o, ℓ^1, L^1 (on a measure space not consisting only of finitely many atoms) two closed, convex, bounded, disjoint sets which cannot be strictly separated. (Hint: Look in the proof of 7.7.)

7.24. Find two closed convex sets in $\mathbb{R}^2$ which cannot be strictly separated.

7.25. How does the proof of 7.7. simplify if the space has the property that each two disjoint closed convex sets, one of which is bounded, can be strictly separated?

7.26. Show that the range of a vector-measure $m: \mathfrak{a} \to E$ is weakly relatively compact, if $\overline{co\ m(\mathfrak{a})}$ is $\mu(E,E')$-complete.

7.27. For every vector-measure $m: \mathfrak{a} \to \mathcal{D}(\Omega)$, Ω open $\subset \mathbb{R}^n$, there is a compact subset $K \subset \Omega$ such that $supp(m(A)) \subset K$ for all $A \subset \mathfrak{a}$. (Hint: Use that every bounded subset of a strict (LF)-space has this property.)

7.28. If $\mathfrak{a}$ is a σ-ring then $m: \mathfrak{a} \to E$, E locally convex, is a vector-measure (with respect to the weak-topology) if and only if $\varphi \circ m$ is σ-additive for all $\varphi \in E'$.

7.29. If $m: \mathfrak{B} \to E$ is a measure on the Borel-sets of a compact set K with values in a quasicomplete space, then there is for every $f \in C(K)$ a unique $x \in E$ with $\int_K f(t)\, \varphi \circ m(dt) = \langle \varphi, x\rangle$. Define therefore

$$x := \int_K f(t)\, m(dt) .$$

(Hint: Use that for finitely many $A_i \in \mathcal{B}$ disjoint, $0 \leq \lambda_i \leq 1$

$$\int \Sigma \lambda_i\, \chi_{A_i}\, dm := \Sigma \lambda_i\, m(A_i) \in \operatorname{co} m(\mathcal{B})$$

and extend continuously. Note that it is enough to assume that the convex hull of the range of m is weakly relatively compact.)

7.30. (a) A weakly compact (linear) operator $T: F \to E$, F a Banach-space, E locally convex, has the property that $T''(F'') \subset E$. (Hint: The $\sigma(F'',F')$-closure of the unit ball of F is the unit ball of F''.)

(b) If K is a compact set, E locally convex, and $T: C(K) \to E$ a weakly compact operator, then there is an E-valued measure m on the Borel-sets of K (with respect to the weak topology of E) such that

$$\int_K f(t)\, m(dt) = Tf$$

for all $f \in C(K)$. (Hint: For a Borel-set A define $m(A) := T''\chi_A \in E$, use 7.28. and the Riesz-representation theorem for $(C(K))'$.)

7.31. (D.P. Milman) A Banach-space E is called uniformly convex if for all $\varepsilon > 0$ there is a $\delta > 0$ such that

$$\left\| \frac{x+y}{2} \right\| \leq 1 - \delta$$

for all $x,y \in E$ with $\|x\| = \|y\| = 1$ and $\|x - y\| \geq \varepsilon$. Show with the aid of James' theorem, that uniformly convex Banach spaces are reflexive.

7.32. Show that

$$T(\xi_n) := ((1 - \|x\|)^{\frac{1}{2}}, \xi_1, \xi_2, \ldots)$$

maps the closed unit ball of ℓ^2 norm-continuously in itself and has no fixed-point. Is T continuous with respect to the weak topology?

§8 THE TOPOLOGY RELATED TO RAINWATER'S THEOREM

8.1. A theorem of J. Rainwater states that a bounded sequence (x_n) in a normed space E converges weakly to zero, if (and only if)

$$\langle \psi, x_n \rangle \to 0$$

for all extreme points ψ of the ($\sigma(E',E)$-compact) dual unit ball U^o. In the framework of Choquet's representation theory this is easy to prove: for every $\varphi \in U^o$ there is a finite non-negative measure μ on ext U^o such that

$$\langle \varphi, x \rangle = \int_{\text{ext } U^o} \langle \psi, x \rangle \mu(d\psi)$$

for all $x \in E$. Rainwater's theorem is therefore an immediate consequence of Lebesgue's dominated convergence theorem. Defining $F := \text{span ext } U^o$ the result reads: a bounded sequence in E converges with respect to $\sigma(E,F)$ if and only if it $\sigma(E,E')$-converges. That the relationship between $\sigma(E,F)$ and $\sigma(E,E')$ is actually much closer will now be investigated by use of James double-limit-theorem: the starting point is that for every $x \in E$

$$x\big|_{U^o} \in \ell^\infty(U^o)$$

is continuous and linear on the compact convex set U^o and attains therefore its supremum in an extreme point of U^o (see 0.7). The basic idea, applying James' theorem to this issue, is due to I. Tweddle. His results were extended by M. De Wilde.

8.2. Let E be a locally convex space, $\mathfrak{U}$ a set of neighbourhoods of zero such that

$$\{\varepsilon U \mid \varepsilon > 0, U \in \mathfrak{U}\}$$

is a basis of neighbourhoods, then all U^o are $\sigma(E',E)$-compact by the theorem of Alaoglu -Bourbaki. By the Krein-Milman theorem U^o is the closed convex hull of its extreme points

$$U^o = \overline{\text{co ext } U^o}^{\,\sigma(E',E)}$$

and therefore Bauer's maximum principle (0.7.) implies that

$$F: = \operatorname{span} \bigcup_{U \in \mathfrak{U}} \operatorname{ext} U^{o}$$

has the property:

(R) For every $U \in \mathfrak{U}$ and $x \in E$,

$$\max_{\varphi \in U^{o}} \langle \varphi, x \rangle = \max_{\varphi \in U^{o} \cap F} \langle \varphi, x \rangle .$$

Exactly this will be used in the sequel. Note, that a subspace $F \subset E$ with (R) is $\sigma(E',E)$-dense in E' (Ex. 8.15.) hence $\sigma(E,F)$ is a Hausdorff-topology on

The fundamental result is the following

THEOREM: Let (E,τ) be a locally convex space, $F \subset E'$ a subspace with the property (R), and $A \subset E$, convex and $\sigma(E,F)$-relatively countably compact. Then

$$\bar{A}^{\sigma(E,F)}$$

is τ-bounded and the topologies $\sigma(E,F)$ and $\sigma(E,E')$ coincide on it.

Proof: With $X: = U^{o}$, $Y: = U^{o} \cap F$, $L: = E|_{U^{o}}$, $Q: = A|_{U^{o}}$

$$A|_{U^{o}} \subset E|_{U^{o}} \hookrightarrow \ell^{\infty}(U^{o})$$

this is precisely the situation described by Corollary 3 of 6.11.: Therefore (with the results of 6.10. (1)-(3)) the topologies $\omega_{U^{o}}$ and $\omega_{U^{o} \cap F}$ coincide on the $\omega_{U^{o}}$-compact $= \omega_{U^{o} \cap F}$-compact set

$$\overline{A|_{U^{o}}}^{\omega_{U^{o}}} = \overline{A|_{U^{o}}}^{\omega_{U^{o} \cap F}}$$

and this set is uniformly bounded on U^{o} (which implies that A is τ-bounded). Now

$$\bar{A}^{\sigma(E,F)} \hookrightarrow \prod_{U \in \mathfrak{U}} \overline{A|_{U^{o}}}^{\omega_{U^{o} \cap F}}$$

and $\Pi\omega_{U^o} = \Pi\omega_{U^o \cap F}$ on the product-space: but these two topologies induce $\sigma(E,E')$ and $\sigma(E,F)$ respectively on $\bar{A}$. ∎

8.3. In the following corollaries E and F are as in the theorem.

COROLLARY 1 (M. De Wilde): Let A be a convex subset of E.

(1) A is $\sigma(E,F)$-compact if and only if it is $\sigma(E,E')$-compact. The same holds for countably compact, sequentially compact, and the relative notions.

(2) If $\bar{A}$ is $\mu(E,E')$-complete and $\sigma(E,F)$-relatively countably compact then it is $\sigma(E,E')$-relatively compact.

Proof: (1) is immediate by the theorem. If $\bar{A}$ is $\mu(E,E')$-complete, the Eberlein-Grothendieck-theorem says that A is $\sigma(E,E')$-relatively compact if it is $\sigma(E,E')$-relatively countably compact. Therefore (1) implies (2). ∎

COROLLARY 2 (I. Tweddle): If E is additionally sequentially complete (x_n) a $\sigma(E,E')$-bounded, $\sigma(E,F)$-null-sequence, then $\{x_n\}$ is contained in a $\sigma(E,E')$-compact absolutely convex subset of E.

Proof: Since $\{x_n\}$ is $\sigma(E,E') = \tau$-bounded and E is sequentially complete

$$T: \ell^1 \longrightarrow E$$

$$(\xi_n) \rightsquigarrow \sum_{n=1}^{\infty} \xi_n x_n$$

is well-defined. Since for every $\varphi \in F$

$$S\varphi := (\langle \varphi, x_n \rangle)_n \in c_o$$

and $c_o' = \ell^1$

$$\langle \varphi, T(\xi_n) \rangle_{F,E} = \sum_n \xi_n \langle \varphi, x_n \rangle = \langle S\varphi, (\xi_n) \rangle_{c_o, \ell^1} ,$$

the operators $T: \ell^1 \to E$ and $S: F \to c_o$ are conjugate and therefore T is $\sigma(\ell^1, c_o)$-$\sigma(E,F)$-continuous. In particular the image TB of the $\sigma(\ell^1, c_o)$-compact unit ball B of ℓ^1 is $\sigma(E,F)$-compact. By Corollary 1 TB is $\sigma(E,E')$-compact. ∎

For a not necessarily sequentially complete space, Corollary 2 applies to the τ-completion (and the same F), therefore (recall Ex. 4.25. on Cauchy-sequences)

COROLLARY 3: The τ-bounded (Cauchy-)sequences are the same for $\sigma(E,F)$ and $\sigma(E,E')$.

In the case that F comes from the extreme points, this is just Rainwater's theorem. It is not completely by chance that this result can be proven with either the aid of James' double-limit-theorem or the theorem of Choquet-Bishop-deLeeuw. S. Simons pointed out that there is a connection between both.

COROLLARY 4: τ-bounded subsets are $\sigma(E,F)$-(relatively) sequentially compact if and only if they are $\sigma(E,E')$-(relatively) sequentially compact.

8.4. Consider the special case $E = C(X)$, X compact, with its sup-norm-topology. Then every extreme point of the dual unit ball is $\pm\delta_x$, the Dirac-measure concentrated in $x \in X$, and therefore $\sigma(C(X),F)$ is the topology ω_X of pointwise convergence. Remembering that both, the pointwise and the weak topology are angelic in this case, Corollary 4 above says the same as Grothendieck's theorem 4.2. .

8.5. For $E = L^1(X,\mathcal{A},\mu) = L^1$, μ a non-negative measure on X such that $(L^1)' = L^\infty$ (this is true for Radon-measures and σ-finite measures) the extreme points of the unit ball of the dual L^∞ are those measurable functions $f: X \to \mathbb{R}$ with $|f(x)| = 1$ almost everywhere or (real functions!) $f = \chi_A - \chi_{\complement A}$ for some $A \in \mathcal{A}$. Since

$$\chi_A = \frac{1}{2}(\chi_X + (\chi_A - \chi_{\complement A}))$$

the span of the extreme points is the span of the characteristic functions that is the space M_o of measurable functions which attain only finitely many values. According to Rainwater's theorem, a bounded sequence (f_n) in L^1 converges to f with respect to $\sigma(L^1,L^\infty)$ if and only if

$$\int_A (f_n - f)d\mu \to o$$

for all $A \in \mathcal{A}$. The requirement that -a priori - the sequence need to be bounded will turn out to be superfluous (8.8.). Theorem 8.2. and its consequences describe some relations between the $\sigma(L^1,L^\infty)$-and $\sigma(L^1,M_o)$-topology. $\sigma(L^1,L^\infty)$ as a weak

topology of a normed space is angelic. It will turn out that $\sigma(L^1, M_o)$ also is angelic; but to see this a careful analysis of the properties of $\sigma(L^1, M_o)$-convergent sequences is necessary. Additionally, these results will lead to a characterization of weak compactness in L^1 in terms of integration properties. (By the way, M. Valdivia has used topological properties of the normed space M_o to study the space of bounded finite measures on a σ-algebra, see [58]).
For this, first the

LEMMA: For $f \in L^1$, and $\varepsilon > 0$ there is a $\delta > 0$ such that

$$\int_D |f| d\mu \leq \varepsilon$$

whenever $\mu(D) \leq \delta$.

Proof: With

$$A_n := [|f| \geq n] \quad n \in \mathbb{N} \quad ,$$

$|f|\chi_{A_n} \to 0$ almost everywhere, hence there is an n with

$$\int_{A_n} |f| d\mu \leq \tfrac{1}{2}\varepsilon$$

If $\mu(D) \leq (2n)^{-1}\varepsilon$, then

$$\int_D |f| d\mu \leq \int_{D \cap A_n} + \int_{D \cap \complement A_n} \leq \tfrac{1}{2}\varepsilon + n\mu(D) \leq \varepsilon \quad . \quad \blacksquare$$

8.6. $\sigma(L^1, M_o)$-convergent sequences have the property, that the statement of the lemma above holds uniformly for them. This will cause a special characterization of weakly compact sets in L^1.

DEFINITION: A subset $H \subset L^1$ is called <u>uniformly integrable</u>, if for every $\varepsilon > 0$ there is a $\delta > 0$ such that for all integrable sets D with $\mu(D) \leq \delta$

$$\int_D |f| d\mu \leq \varepsilon$$

holds for all $f \in H$.

By passing to positive and negative parts, it is enough to check

$$\left| \int_D f \, d\mu \right| \leq \varepsilon$$

for uniform integrability. The lemma 8.5. stated that one-point (and therefore all finite) subsets of L^1 are uniformly integrable. It follows immediately that lattice-bounded sets

$$\{h \in L^1 \mid |h| \leq f\}$$

$f \in L^1$, are uniformly integrable.

PROPOSITION: If (f_n) is a $\sigma(L^1, M_0)$-Cauchy-sequence then $\{f_n\}$ is uniformly integrable.

Proof: For $S := \{D \subset X \mid D \text{ integrable}\} \subset L^1 \cap L^\infty$ via characteristic functions (usual identification) the notations

$$\langle D, g\rangle := \int_D g\, d\mu \qquad D \in S,\ g \in L^1$$

$$d(B,D) := \int |\chi_B - \chi_D|\, d\mu \qquad B, D \in S$$

are convenient. d is a metric on S.

(a) If $g \in L^1$, $D_0 \in S$, $\epsilon > 0$, and $\delta > 0$, $|\langle D_0, g\rangle| \leq \epsilon$, and $|\langle B, g\rangle| \leq \epsilon$ for all $B \in S$ with $d(B, D_0) \leq \delta$, then for all $D \in S$ with $\mu(D) \leq \delta$

$$|\langle D, g\rangle| \leq 4\epsilon \quad .$$

To prove this, assume first that $D \subset D_0$ and $\mu(D) \leq \delta$. Then

$$\chi_D = \chi_{D_o} - \chi_{D_0 \setminus D}$$

therefore $d(D_0, D_0 \setminus D) = \mu(D) \leq \delta$ and

$$|\langle D, g\rangle| \leq |\langle D_0, g\rangle| + |\langle D_0 \setminus D, g\rangle| \leq 2\epsilon$$

If $D \subset X \setminus D_0$, $\mu(D) \leq \delta$, then

$$\chi_D = \chi_{D \cup D_o} - \chi_{D_o}$$

and the same argument holds. Naturally, for an arbitrary A, a partition gives the desired result.

(b) Since every L^1-convergent sequence has an almost everywhere convergent subsequence, (S,d) is a complete metric space - and, by the same argument, for every $g \in L^1$ and $\varepsilon > 0$, the set

$$S(g,\varepsilon) := \{B \in S \mid |\langle B,g\rangle| \le \varepsilon\}$$

is closed.

(c) Now, by the assumption, that $(\langle B,f_n\rangle)_n$ is a Cauchy-sequence for every $B \in S$, it follows that, for a given $\varepsilon > 0$,

$$S = \bigcup_{k=1}^{\infty} \bigcap_{m,n \ge k} S(f_m - f_n, \varepsilon)$$

and the Baire category theorem ascertains the existence of $k_0 \in \mathbb{N}$, $D_0 \in S$, and $\delta > 0$ such that

$$B \in S(f_n - f_m, \varepsilon) \qquad m,n \ge k_0$$

whenever $d(B,D_0) \le \delta$. Moreover, by the convergence of $(\langle D_0,f_n\rangle)$ it can be assumed, that $|\langle D_0,f_n - f_m\rangle| \le \varepsilon$ for $n,m \ge k_1 \ge k_0$: Altogether, this is just the situation of (a) for all $g := f_n - f_m$, so

$$|\langle D, f_n - f_m\rangle| \le 4\varepsilon$$

whenever $\mu(D) \le \delta$ and $n,m \ge k_1$. The finite set $\{f_\ell \mid \ell \le k_1\}$ is certainly uniformly integrable, so there is a $\delta_1 < \delta$ such that

$$|\langle D,f_\ell\rangle| \le \varepsilon \qquad \ell \le k_1$$

for D with $\mu(D) \le \delta_1$. Hence, if $\mu(D) \le \delta_1$ and $n \ge k_1$

$$|\langle D,f_n\rangle| \le |\langle D,f_n - f_{k_1}\rangle| + |\langle D,f_{k_1}\rangle| \le 5\varepsilon :$$

The set $\{f_n\}$ is uniformly integrable. ■

8.7. The proposition says that $\sigma(L^1,M_0)$ - convergent sequences are uniformly small for small sets. However, a trick used by A. Grothendieck will show that they are also small off integrable sets:

COROLLARY 1: If (f_n) is a $\sigma(L^1, M_o)$-Cauchy-sequence, then for every $\varepsilon > 0$ there is an integrable set K, such that

$$\int_{X\setminus K} |f_n| d\mu \le \varepsilon$$

for all $n \in \mathbb{N}$.

Proof. Define

$$g = \sum_{n=1}^{\infty} \frac{1}{2^n} \frac{|f_n|}{\|f_n\|_1} \in L^1(\mu) \qquad \left(\frac{o}{o} := o\right) .$$

Then $\nu := g\mu$ is a finite measure on $(X, \mathcal{A})$ and

$$g_n := \frac{f_n}{g} \in L^1(\nu) .$$

For $B \in \mathcal{A}$

$$\int_B g_n \, d\nu = \int_B f_n \, d\mu$$

and therefore the proposition applies to the sequence (g_n) in $L^1(\nu)$: For given $\varepsilon > 0$, there is a $\delta > 0$ such that

$$\varepsilon \ge \int_C |g_n| d\nu = \int_C \frac{|f_n|}{g} g d\mu = \int_C |f_n| d\mu$$

for all C with

$$\nu(C) = \int_C g \, d\mu \le \delta .$$

Since there are integrable sets $D_n \subset D_{n+1} \ldots$ such that

$$\lim_n \int_{D_n} g \, d\mu = \int g \, d\mu$$

there is a $K := D_n$ with $\int_{X\setminus D_n} g \, d\mu \le \delta$. ■

A special case is the space ℓ^1 of summable sequences - m_o denoting those sequences which attain only finitely many values.

COROLLARY 2: For a sequence (x_n) in ℓ^1 the following statements are equivalent:

(1) (x_n) is $\sigma(\ell^1, m_o)$-convergent.

(2) (x_n) is $\sigma(\ell^1, \ell^\infty)$-convergent.

(3) (x_n) is norm-convergent.

The same holds for Cauchy-sequences.

This result is known as Schur's lemma, which can also be proved directly with the method of a gliding hump.

Proof: Only (1)$\rightsquigarrow$(3) is to check for a null-sequence (x_n): For $\epsilon > 0$ there is, according to Corollary 1, an index k_o such that for all n

$$\sum_{k=k_o}^{\infty} |x_{n,k}| \leq \epsilon \quad .$$

This, together with the coordinate-wise convergence, implies norm-convergence. As always (Ex. 4.25.) if two locally convex topologies have the same convergent sequences, they have the same Cauchy-sequences. ■

In particular, ℓ^1 is $\sigma(\ell^1, m_o)$- and $\sigma(\ell^1, \ell^\infty)$-sequentially complete and (relative) sequential compactness is the same for $\sigma(\ell^1, m_o)$, $\sigma(\ell^1, \ell^\infty)$, and the norm-topology! Furthermore, ℓ^1 is weakly sequentially complete, but not weakly quasicomplete.

8.8. In particular these sequences were bounded. This is generally true:

COROLLARY 3: $\sigma(L^1, M_o)$-Cauchy-sequences are norm-bounded.

Proof: According to Corollary 1, there is an integrable set K such that

$$\int_{X\setminus K} |f_n| d\mu \leq 1 \quad \text{for all} \quad n \in \mathbb{N}$$

and the uniform integrability gives a $\delta > 0$ such that

$$\int_D |f_n| d\mu \leq 1 \quad \text{for all} \quad n \in \mathbb{N}$$

whenever $\mu(D) \leq \delta$: The integrable set K can be decomposed

$$K = \bigcup_{m=1}^{m_0} D_m$$

where, for all m, either $\mu(D_m) \leq \delta$ or D_m is an atom. Since (f_n) converges pointwise on atoms, these observations yield immediately that $\{f_n\}$ is norm-bounded. ■

This removes the boundedness requirement for convergent sequences in 8.5. .

COROLLARY 4: Every $\sigma(L^1,M_0)$-bounded subset $H \subset L^1$ is norm-bounded.

Proof: Assume there is a sequence (x_n) in H with $\|x_n\| \geq n^2$. Since H is $\sigma(L^1,M_0)$-bounded the sequence $(n^{-1}x_n)$ is $\sigma(L^1,M_0)$-convergent to zero and therefore, by Corollary 3, norm-bounded. This is a contradiction. ■

P. Dierolf remarked that this implies the

COROLLARY 5:

(a) $\sigma(L^1,L^\infty)$ and $\sigma(L^1,M_0)$ coincide on $\sigma(L^1,M_0)$-bounded (= norm-bounded) subsets of L^1 .

(b) $\sigma(L^1,M_0)$ is angelic.

Proof: By the definition of an angelic locally convex space the following is easy to check:

> Let E be a vector space with locally convex topologies τ_1 and τ_2 which coincide on τ_2-bounded subsets of E. If τ_1 is angelic and finer than τ_2 , then τ_2 is also angelic.

Since $\sigma(L^1,L^\infty)$ is angelic, this remark and Corollary 4 imply that it is enough to show that on norm-bounded subsets $H \subset L^1$ the topologies $\sigma(L^1,M_0)$ and $\sigma(L^1,L^\infty)$ coincide.

For this, observe that M_o is norm-dense in L^∞ hence $\sigma((L^\infty)',M_o)$ is a Hausdorff-topology on $(L^\infty)'$ coarser than $\sigma((L^\infty)',L^\infty)$. Therefore $\sigma((L^\infty)',L^\infty) = \sigma((L^\infty)',M_o)$ on $\sigma((L^\infty)',L^\infty)$-relatively compact = norm-bounded subsets of $(L^\infty)'$: In particular $\sigma(L^1,M_o) = \sigma(L^1,L^\infty)$ on norm-bounded subsets of L^1. ■

Note that this Corollary implies that the $\sigma(L^1,M_o)$- and the $\sigma(L^1,L^\infty)$-closure of a $\sigma(L^1,M_o)$-relatively countably compact subset H of L^1 coincide and every $f \in \bar{H}$ is the $\sigma(L^1,L^\infty)$-limit of a sequence in H.

8.9. After these preparations the main result on weak compactness in L^1 can be formulated:

THEOREM (N. Dunford - B.J. Pettis): For a subset $H \subset L^1$ the following statements are equivalent:

(1) H is $\sigma(L^1,M_o)$-relatively compact (= rel. sequentially compact = rel. countably compact).

(2) H is $\sigma(L^1,L^\infty)$-relatively compact (= rel. sequentially compact = rel. countably compact).

(3) H is

(a) norm-bounded

(b) uniformly integrable

(c) for every $\varepsilon > 0$ there is an integrable set $K \subset X$ such that

$$\int_{X\setminus K} |f|\,d\mu \le \varepsilon$$

for all $f \in H$.

In finite measure spaces (3c) is redundant - if, moreover, there are no atoms, then also (3a) can be deleted (see Ex. 8.20.). In the case of the sequence space ℓ^1 the condition (3b) is empty; also (relative) weak compactness is the same as (relative) norm-compactness by Schur's lemma 8.7. .

The present approach to the theorem of Dunford-Pettis follows essentially the exposition [11] of J. Dieudonné.

Proof:

(2)⟺(1) is just Corollary 5 .

(1)⟹(3) If H were not bounded, there were a sequence (f_n) in H, which is unbounded: Passing to a $\sigma(L^1, M_o)$-convergent subsequence, this would violate that, by 8.8., all these sequences are bounded. Similarly, (3b) holds by Proposition 8.6. Assume that (3c) is not true: by induction there are integrable $K_n \subset X$ and $f_n \in H$ with

$$\int_{X \setminus K_n} |f_n| d\mu \geq \epsilon \quad \text{and} \quad \int_{X \setminus K_n} |f_m| d\mu \leq \frac{1}{n} \quad \text{for} \quad m < n$$

Assuming, again by the sequential compactness, that (f_n) is $\sigma(L^1, M_o)$-convergent, Corollary 1 in 8.7. applied to $X_o := \bigcup_{n=1}^{\infty} K_n$, gives the existence of an integrable $K \subset X_o$ with

$$\int_{X_o \setminus K} |f_n| d\mu \leq \frac{1}{3}\epsilon$$

for all n. Since all f_n vanish off X_o

$$\epsilon \leq \int_{X \setminus K_n} |f_n| d\mu \leq \int_{X_o \setminus K} |f_n| d\mu + \int_{K \cap (X_o \setminus K_n)} |f_n| d\mu \leq \frac{1}{3}\epsilon + \frac{1}{3}\epsilon$$

- the latter for large n, since $\{f_n\}$ is uniformly integrable and $\mu(K \cap (X_o \setminus K_n)) \to 0$.

(3)⟹(2): Since H is bounded

$$\overline{H}^{\sigma((L^\infty)', L^\infty)}$$

is $\sigma((L^\infty)', L^\infty)$-compact, and it is therefore enough to show, that each $\varphi \in \overline{H}$ is already in L^1, that is: φ is represented by an integrable function.

The set function

$$\nu(A) := \langle \varphi, \chi_A \rangle$$

is finitely additive on $\mathfrak{a}$ and, according to (3b), there is for every $\epsilon > 0$ a $\delta > 0$ such that

$$(*) \qquad |\nu(D)| \leq \sup_{f \in H} \left| \int_D f \, d\mu \right| \leq \epsilon$$

if $\mu(D) \leq \delta$. Furthermore (3c) produces a sequence $K_m \subset K_{m+1}$ of integrable sets such that for all measurable $B \subset X \setminus K_m$

$$|\nu(B)| \leq \sup_{f \in H} \int_B |f| d\mu \leq \sup_{f \in H} \int_{X \setminus K_n} |f| d\mu \leq \frac{1}{m} \quad .$$

To show, that ν is actually σ-additive it is enough to prove $\nu(A_n) \to 0$ whenever $A_n \in \mathcal{A}$ and $A_n \downarrow \emptyset$: If $\mu(A_n) < \infty$ then $\mu(A_n) \downarrow 0$ and hence $\nu(A_n) \to 0$ by (*). But this means, that in the general case

$$\nu(A_n) = \nu(A_n \cap K_m) + \nu(A_n \cap \complement K_m) \leq \frac{2}{m}$$

for $n \geq n_0$.

Since ν is μ-absolutely-continuous (again by (*)), the Radon-Nikodym-theorem delivers a locally integrable function f_0 such that

$$(**) \qquad \int_D f_0 d\mu = \nu(D)$$

for all μ-integrable $D \in \mathcal{A}$ and whose support is in $X_0 := \bigcup_m K_m$. Since ν is finite, the Beppo-Levi-theorem (taking positive and negative parts) implies that $f_0 \in L^1(\mu)$. Hence, by the σ-additivity of both sides of (**), this equation holds for all $A \in \mathcal{A}$, (first only for $A \subset X_0$ - but outside X_0 both, f_0 and ν, are zero). This means that

$$\int g f_0 \, d\mu = \langle \varphi, g \rangle$$

for all $g \in M_0 \subset L^\infty$: but M_0 is norm-dense in L^∞ and f_0 and φ, considered as functionals on L^∞ are continuous; so they are equal in $(L^\infty)'$, hence $\varphi \in L^1$. ■

The content of the last step of the proof will be examined in Ex. 8.23..

COROLLARY: L^1 is $\sigma(L^1,M_0)$ and $\sigma(L^1,L^\infty)$ - sequentially complete.

Proof: According to the results in 8.6.-8.8. a $\sigma(L^1,M_0)$ Cauchy-sequence (f_n) satisfies all of the requirements in (3) of the Dunford-Pettis theorem; consequently, the set $\{f_n\}$ is $\sigma(L^1,M_0) = \sigma(L^1,L^\infty)$-relatively sequentially compact and (f_n) converges therefore with respect to these topologies. ■

8.10. Another situation where the "extreme-point"-topology $\sigma(E,F)$ from the beginning of this paragraph produces some results, is the ε-tensor product of spaces, which is mostly used to describe spaces of vector-valued functions.

For locally convex spaces E_1 and E_2 the 4-linear map

$$\begin{array}{ccc} E_1 \times E_1' \times E_2 \times E_2' & \longrightarrow & \mathbb{R} \\ \cup\!\!\!\shortmid & & \cup\!\!\!\shortmid \\ (\varphi, x, \psi, y) & \rightsquigarrow & \langle \varphi, x\rangle \langle \psi, y\rangle \end{array}$$

defines a natural duality pairing between the tensor products:

$$\langle E_1' \otimes E_2' , E_1 \otimes E_2 \rangle .$$

The ε-topology (or topology of bi-equicontinuous convergence) on $E_1 \otimes E_2$ is given by the semi-norms

$$E_1 \otimes E_2 \ni z \rightsquigarrow \sup_{\varphi \in U_1^o,\ \psi \in U_2^o} |\langle \varphi \otimes \psi , z\rangle|$$

$U_i \in \mathfrak{U}_i$, $i = 1,2$, where $\mathfrak{U}_i$ are bases of neighbourhoods of E_i as in 8.7. $E_1 \otimes_\varepsilon E_2$ denotes $E_1 \otimes E_2$ equipped with this topology, $E_1 \tilde{\otimes}_\varepsilon E_2$ the completion. It is obvious that

$$E_1' \otimes E_2' \subset (E_1 \tilde{\otimes}_\varepsilon E_2)'$$

with respect to the natural pairing - and, less obvious, that for every $\Phi \in (E \tilde{\otimes}_\varepsilon E_2)'$ there are $U_i \in \mathfrak{U}_i$ and a non-negative Radon-measure on the compact (weak topologies) set $U_1^o \times U_2^o$ such that

$$\langle \Phi, z\rangle = \int_{U_1^o \times U_2^o} \langle \varphi \otimes \psi , z\rangle \mu(d(\varphi,\psi))$$

for every $z \in E_1 \tilde{\otimes}_\varepsilon E_2$. Lebesgue's dominated convergence theorem gives immediately that weak convergence of bounded sequences (z_n) in $E_1 \otimes_\varepsilon E_2$ is the same as the $\sigma(E_1 \otimes E_2 , E_1' \otimes E_2')$-convergence.

8.11. The latter statement can also be shown (and extended considerably), without using the representation (*), in the present framework of Rainwater's theorem: A typical equicontinuous set in $(E_1 \otimes_\varepsilon E_2)'$ has the form

$$U^o = (U_1^o \otimes U_2^o)^{oo} = \overline{\text{co}\, U_1^o \otimes U_2^o}$$

by the bipolar theorem - the bipolar taken in $\langle (E_1 \otimes_\varepsilon E_2)', E_1 \otimes E_2 \rangle$. Since the bilinear map

$$\otimes: (E_1', \sigma(E_1', E_1)) \times (E_2', \sigma(E_2', E_2)) \rightarrow [(E_1 \otimes_\varepsilon E_2)', \sigma((E_1 \otimes_\varepsilon E_2)', E_1 \otimes E_2)]$$

is certainly continuous, the following lemma can be applied.

LEMMA: Let H_1, H_2, G be locally convex spaces $K_i \subset H_i$ compact and convex, $T: H_1 \times H_2 \rightarrow G$ a continuous bilinear map then

$$\text{ext}\, \overline{\text{co}\, T(K_1 \times K_2)} \subset T(\text{ext}\, K_1 \times \text{ext}\, K_2) .$$

Proof:

(a) $\text{ext}\, K_1 \times \text{ext}\, K_2 = \text{ext}(K_1 \times K_2)$:
If $(x_o, y_o) \in \text{ext}\, K_1 \times \text{ext}\, K_2$ and (for $(x_i, y_i) \in K_1 \times K_2$, $\alpha + \beta = 1$, $\alpha > 0, \beta > 0$)

$$(x_o, y_o) = \alpha(x_1, y_1) + \beta(x_2, y_2)$$

then $x_o = \alpha x_1 + \beta x_2$ and hence $x_o = x_1 = x_2$; the same for the other coordinate.

Conversely, if $(x_o, y_o) \in \text{ext}(K_1 \times K_2)$ and $x_o = \alpha x_1 + \beta x_2$, then $(x_o, y_o) = \alpha(x_1, y_o) + \beta(x_2, y_o)$ and therefore $(x_1, y_o) = (x_2, y_o)$.

(b) For every $z \in \text{ext}\, \overline{\text{co}\, T(K_1 \times K_2)}$ the set $D := T^{-1}\{z\} \cap (K_1 \times K_2)$ is non-empty and extremal (see 0.7.) in $K_1 \times K_2$:

First, $T(K_1 \times K_2)$ is compact, such that the Milman theorem ascertains that all extreme points of $\overline{\text{co}\, T(K_1 \times K_2)}$ are in $T(K_1 \times K_2)$; hence $z \in T(K_1 \times K_2)$ and $D \neq \emptyset$.

To show that D is extremal take $(x_i, y_i) \in K_1 \times K_2$, $i = 0,1,2$, $T(x_o, y_o) = z$,

$$\alpha(x_1, y_1) + \beta(x_2, y_2) = (x_o, y_o) .$$

Then

$$z = T(x_o, y_o) = T(x_o, \alpha y_1 + \beta y_2) = \alpha T(x_o, y_1) + \beta T(x_o, y_2)$$

and therefore $z = T(x_0,y_0) = T(x_0,y_1) = T(x_0,y_2)$. The same argument applied to

$$z = T(x_0,y_1) = T(\alpha x_1 + \beta x_2, y_1) = \dots$$

yields $z = T(x_1,y_1) = T(x_2,y_1)$, and, again the same, $z = T(x_2,y_2)$.

(c) Now the proof can be finished: For $z \in \operatorname{ext} \overline{\operatorname{co} T(K_1 \times K_2)}$ the pre-image $T^{-1}\{z\} \cap (K_1 \times K_2)$ is by (b) extremal, closed, and non-empty in the compact set $K_1 \times K_2$, contains therefore an extreme point (0.7.)

$$(x_0,y_0) \in \operatorname{ext} K_1 \times K_2 = \operatorname{ext} K_1 \times \operatorname{ext} K_2$$

(by (a)), hence $T(x_0,y_0) = z$. ∎

8.12. Returning to $U^0 = \overline{\operatorname{co} U_1^0 \otimes U_2^0}$, this lemma says that

$$\operatorname{ext} U^0 \subset \operatorname{ext} U_1^0 \otimes \operatorname{ext} U_2^0$$

and hence

$$F_0 := \operatorname{span} \bigcup_{U_i \in \mathfrak{U}_i} \operatorname{ext} U_1^0 \otimes \operatorname{ext} U_2^0 \subset (E_1 \otimes_\varepsilon E_2)'$$

satisfies the property (R) of 8.2. for $E = E_1 \tilde{\otimes}_\varepsilon E_2$: Theorem 8.2. and its corollaries 1 - 4 apply. Certainly all vector-spaces F between F_0 and $(E_1 \otimes_\varepsilon E_2)'$ also fulfill (R) , in particular the case $F := E_1' \otimes E_2'$ includes the above-mentioned characterization of convergent sequences and generalizes some results obtained by L. Tsitsas.

8.13. If X is compact, E a complete locally convex space, then the space of continuous E-valued functions on X is the completed ε-tensor product

$$C(X,E) = C(X) \tilde{\otimes}_\varepsilon E$$

- the ε-topology is the topology of uniform convergence, and the semi-norms are (p a continuous semi-norm on E)

$$\sup_{x \in X} p(f(x)) \quad .$$

Since the extreme points of the unit ball in $C(X)'$ are just $\pm \delta_x$, the vector-space

$$F := \operatorname{span} X \otimes E' \subset C(X,E)'$$

is a space as in 8.12.: the topology $\sigma(C(X,E), F)$ is the topology of pointwise convergence of functions $X \to (E, \sigma(E,E'))$.

If F_o (as in 8.12.) is chosen to be the vector-space

$$F_o := \operatorname{span} X \otimes \operatorname{span} \bigcup_{U \in \mathfrak{U}_E} \operatorname{ext} U^o$$

the associated weak topology may be coarser than the former. Take for example $E = \mathcal{E}(\Omega)$, the space of infinitely often differentiable functions on an open subset $\Omega \subset \mathbb{R}^N$ with the topology of uniform convergence of all derivatives on all compact sets $K \subset \Omega$. With $(\alpha \in (\mathbb{N} \cup \{0\})^N)$

$$U_{\alpha,K} := \{f \mid \sup_{\omega \in K} |D^\alpha f(\omega)| \leq 1\}$$

it is easily seen that $(D^\alpha \delta_\omega) f := (-1)^{|\alpha|} D^\alpha f(\omega)$

$$\operatorname{ext} U^o_{\alpha,K} := \{\varepsilon D^\alpha \delta_\omega \mid \varepsilon = \pm 1, \omega \in K\}$$

and, consequently, for a typical neighbourhood of zero

$$U = \bigcap_{|\alpha| \leq m} U_{\alpha,K}$$

the extreme points of

$$U^o = \operatorname{co}\Big(\bigcup_{|\alpha| \leq m} U^o_{\alpha,K}\Big)$$

are $\pm$ derivatives of Dirac functionals. Therefore

$$F_1 := \operatorname{span} X \otimes \operatorname{span} \{\varepsilon D^\alpha \delta_\omega \mid \varepsilon = \pm 1, \omega \in \Omega, \alpha \in (\mathbb{N} \cup \{0\})^N\}$$

and $\sigma(C(X, \mathcal{E}(\Omega)), F_1)$ on $C(X, \mathcal{E}(\Omega))$, considered as a space of functions on $X \times \Omega$, is the topology of pointwise convergence in the first variable and all derivatives in the second variable. Since $C(X, \mathcal{E}(\Omega))$ is complete, Corollary 1 for example says that the convex $\sigma(\ldots, F_1)$-compact sets are weakly compact; Corollary 3, that for a bounded sequence the (Cauchy) convergences coincide.

8.14. Undoubtedly, many of the applications of the results in 8.1-8.3. can be obtained directly, and often easier. But the Rainwater-theorem and the related weak topology give a unified approach.

EXERCISES

8.15. A subspace $F \subset E'$ with (R) is $\sigma(E',E)$-dense.

8.16. Find an example proving that in Rainwater's theorem the boundedness-assumption cannot be deleted.

8.17. Show that for a Hilbert-space E there are no proper subspaces $F \subset E'$ satisfying (R).

8.18. What is the meaning of Rainwater's theorem - and the results of 8.2. and 8.3. - in c_o ?

8.19. Let A be a compact subset of a locally convex space E, (φ_n) in E' a sequence which is uniformly bounded on A, with $\langle\varphi_n,x\rangle \to 0$ for all $x \in A$. Then $\langle\varphi_n,y\rangle \to 0$ for all $y \in \overline{\text{co}\,A}$. (Hint: Milman's theorem 0.7. implies for $B := \overline{\Gamma A}$ that $\text{ext}\,B \subset A \cup (-A)$. For $G = \ker m_{B^o}$, m_{B^o} the Minkowski-functional of B^o, the dual of the normed space $(E'/G, m_{B^o})$ is $[\![B]\!]$. Apply Rainwater's theorem.)

This result, together with the Grothendieck-Eberlein theorem on double-limits can also be used to prove Krein's theorem ([45]).

8.20. (a) Show that in the theorem of Dunford-Pettis the boundedness-condition cannot be deleted.(Hint: Take a space consisting of one atom.)

(b) In finite measure-spaces without atoms uniformly integrable sets are norm-bounded.

8.21. (a) If $H \subset L^1$ is weakly relatively compact, then

$$\{g \in L^1 \mid \exists h \in H \; |g| \le |h|\}$$

is weakly relatively compact.

(b) Lattice-bounded subsets of L^1 are weakly relatively compact.

8.22. Show that Schur's lemma holds also in $\ell'(I)$, I an arbitrary index set.

8.23. An element $\varphi \in (L^\infty)'$ is in L^1 if and only if for every $\varepsilon > 0$ there exist (1) a $\delta > 0$ such that $|\langle \varphi, \chi_D \rangle| \le \varepsilon$ whenever $\mu(D) \le \delta$, and (2) an integrable set K with $|\langle \varphi, \chi_B \rangle| \le \varepsilon$ for all $B \subset X \setminus K$. (Hint: Look into the proof of the Dunford-Pettis theorem.)

8.24. Show that for every $\sigma(L^1, L^\infty)$-relatively compact set H in $L^1(X,\mu)$, there is a σ-finite measurable set $X_o \subset X$ such that all $f \in H$ are zero off X_o.

8.25. In $L^1(X,\mu)$ a sequence (f_n) is said to converge in measure to $f \in L^1$ if for every $\eta > 0$

$$\mu\left[|f_n - f| \ge \eta\right] \to 0$$

Prove, that (f_n) norm-converges if and only if it converges in measure, is uniformly integrable and for every $\varepsilon > 0$ there is an integrable K such that

$$\int_{X\setminus K} |f_n| d\mu \le \varepsilon$$

for all n. (Hint: One implication can be checked directly, for the other one consult 8.6. and 8.8.)

8.26. For an open set $\Omega \subset \mathbb{R}^N$, a sequence (f_n) in $\mathcal{D}(\Omega)$ converges weakly to zero if and only if it vanishes off a common compact set $K \subset \Omega$, $D^\alpha f_n$ converges pointwise to zero and is uniformly bounded for each derivative.

BIBLIOGRAPHY

The following list of papers and books is not intended to be a complete bibliography on weakly compact sets. It contains only items which are related to the text.

[1] Bishop, E. and Phelps, R. R.: A proof that every Banach space is subreflexive, Bull. Amer. Math. Soc., 67 (1961) 97-98.

[2] Boehme, T. K. and Rosenfeld, M.: An example of two compact Hausdorff Fréchet-spaces whose product is not Fréchet, J. London Math. Soc., 8 (1974) 339-344.

[3] Buchwalter, H.: Parties bornées d'un espace topologique complètement régulier, Sém. Choquet, 9 (1969/70) no. 14.

[4] Cellina, A.: On the non-existence of solutions of differential equations in non-reflexive spaces, Bull. Amer. Math. Soc., 72 (1972) 1069-1072.

[5] Constantinescu, C.: Šmulian-Eberlein spaces, Comm. Math. Helv., 48 (1973) 254-317.

[6] Davis, W. J., Figiel, T., Johnson, W. B., and Pełczyński, A.: Factoring weakly compact operators, J. Funct. Analysis, 17 (1974) 311-327.

[7] De Wilde, M.: Pointwise compactness in spaces of functions and weak compactness in locally convex spaces, Sém. Analyse Fonctionnelle et Applications (H. G. Garnir) Liège, 1973.

[8] De Wilde, M.: Pointwise compactness in spaces of functions and R. C. James theorem, Math. Ann., 208 (1974) 33-47.

[9] De Wilde, M.: Doubles limites ordonnés et théorèmes de minimax, Ann. Inst. Fourier, 24, 4 (1974) 181-188.

[10] Dierolf, P.: Theorems of Orlicz-Pettis-type for locally convex spaces, manuscr. math., 20 (1977) 73-94.

[11] Dieudonné, J.: Sur les espaces de Köthe, J. Analyse Math., 1 (1951) 81-115.

[12] Dieudonné, J. and Gomes, A. P.: Sur certains espaces vectoriels topologiques, C.R.A.S.P., 230 (1950) 1129-1130.

[13] Dunford, N. and Pettis, B. J.: Linear operations on summable functions, Trans. Amer. Math. Soc., 47 (1940) 323-392.

[14] Eberlein, W. F.: Weak compactness in Banach spaces, I., Proc. Nat. Acad. Sci. USA, 33 (1947) 51-53.

[15] Edelstein, M. and Thompson, A. C.: Some results on nearest points and support properties of convex sets in c_o, Pacific J. Math., 40 (1972) 553-559.

[16] Edwards, R. E.: Functional analysis: theory and applications, Holt, Rinehart and Winston, 1965.

[17] Floret, K. and Wloka, J.: Einführung in die Theorie der lokalkonvexen Räume, Lect. Notes Math. 56 (1968).

[18] Floret, K. and Wriedt, M.: Reflexivität und Bestapproximation in Fréchet-Räumen, Arch. Math., 23 (1972) 70-72.

[19] Godunov, A. N.: Peano's theorem in Banach spaces, Funct. Anal. Appl., 9 (1975) 53-55.

[20] Govaerts, W.: A productive class of angelic spaces, to appear.

[21] Grothendieck, A.: Critères généraux de compacité dans les espaces vectoriels localement convexes, Pathologie des espaces (LF), C.R.A.S.P., 231 (1950) 940-942.

[22] Grothendieck, A.: Critères de compacité dans les espaces fonctionnels généraux, Amer. J. Math., 74 (1952) 168-186.

[23] Grothendieck, A.: Espaces vectoriels topologiques, Publ. Soc. Mat. Sao Paulo, 1958.

[24] Horvath, J.: Topological vector spaces and distributions, Addison-Wesley, 1966.

[25] James, R. C.: Bases and reflexivity of Banach spaces, Bull. Amer. Math. Soc., 56 (1950) 58.

[26] James, R. C.: Reflexivity and the supremum of linear functionals, Annals of Math., 66 (1957) 159-169.

[27] James, R. C.: Weakly compact sets, Trans. Amer. Math. Soc., 113 (1964) 129-140.

[28] James, R. C.: Weak compactness and reflexivity, Israel J. Math., 2 (1964) 101-119.

[29] James, R. C.: A counterexample for a sup-theorem in normed spaces, Israel J. Math., 9 (1971) 511-512.

[30] James, R. C.: Reflexivity and the sup of linear functionals, Israel J. Math., 13 (1972) 289-300.

[31] Kelley, J. L. and Namioka, I., et al.: Linear topological spaces, Van Nostrand, 1963.

[32] Klee, V.: Some characterizations of reflexivity, Rev. Ci. Lima, 52 (1950) 15-23.

[33] Klee, V.: Some topological properties of convex sets, Trans. Amer. Math. Soc., 78 (1955) 30-45.

[34] Klee, V.: A conjecture on weak compactness, Trans. Amer. Math. Soc., 104 (1962) 398-402.

[35] Kluvanek, I. and Knowles, G.: Vector measures and control systems, North Holland Math. Studies, 20 (1975).

[36] Köthe, G.: Topological vector spaces I, Grundl. Math. Wiss. 159, Springer 1969.

[37] Mazur, S.: Über schwache Konvergenz in L^p-Räumen, Studia Math., 4 (1933) 128-133.

[38] Milman, D. P. and Milman, V. D.: Some geometric properties of nonreflexive spaces, Soviet Math., 4 (1963) 1250-1252.

[39] Phelps, R. R.: Lectures on Choquet's theorem, Van Nostrand Math. Studies, 7 (1966).

[40] Pryce, J. D.: Weak compactness in locally convex spaces, Proc. Amer. Math. Soc., 17 (1966) 148-155.

[41] Pryce, J. D.: A device of R. J. Whitley's applied to pointwise compactness in spaces of continuous functions, Proc. London Math. Soc., 23 (1971) 537-546.

[42] Pták, V.: A combinatorial lemma on the existence of convex means and its applications to weak compactness, Proc. Symp. Pure Math. Amer. Math. Soc., 7 (1963) 437-450.

[43] Raikov, D. A.: The work of V. L. Šmulian on topological linear spaces, Russ. Math. Surveys (U.M.N.) 20, 2 (1965) 130-141.

[44] Rainwater, J.: Weak convergence of bounded sequences, Proc. Amer. Math. Soc., 14 (1963) 999.

[45] Schaefer, H. H.: Topological vector spaces, Springer GTM,3 (1971).

[46] Schmets, J.: Espaces de fonctions continues, Lect. Notes Math., 519 (1976).

[47] Sibony, Daniel: Le théorème de James, Sém. Brelot-Choquet-Deny, (1967/68) no. 4.

[48] Simons, S.: A convergence theorem with boundary, Pacific J. Math., 40 (1972) 703-708.

[49] Simons, S.: Maximinimax, minimax, and antiminimax theorems and a result of R. C. James, Pacific J. Math., 40 (1972) 709-718.

[50] Simons, S.: On Pták's combinatorial lemma, Pacific J. Math., 40 (1972) 719-721.

[51] Šmulian, V.: Über lineare topologische Räume, Mat. Sbornik, 7 (1940) 425-448.

[52] Tsitsas, L.: On the heredity of weak compactness in biprojective tensor product-spaces, Studia Math., 61 (1977) 1-6.

[53] Tweddle, I.: Weak compactness in locally convex spaces, Glasgow Math. J., 9 (1968) 123-127.

[54] Tweddle, I.: Results involving weak compactness in $C(X)$, $C(\upsilon X)$ and $C(X)'$, Proc. Edinburgh Math. Soc., 19 (1974/75) 221-229.

[55] Valdivia, M.: Some criteria for weak compactness, J. reine angew. Math., 225 (1972) 165-169.

[56] Valdivia, M.: On weak compactness, Studia Math., 49 (1973) 35-40.

[57] Valdivia, M.: Some new results on weak compactness, J. Funct. Analysis, 24 (1977) 1-10.

[58] Valdivia, M.: On certain barrelled normed spaces, to appear in Ann. Inst. Fourier.

[59] Wilansky, A.: Topology for analysis, Ginn and Company, 1970.

REFERENCES TO THE SECTIONS

§0 [16],[17],[23],[24],[31],[36],[39],[45]

§1 [7],[21],[22],[23],[31]

§2 [57]

§3 [7],[31],[36],[41]

§4 [3],[5],[7],[8],[46],[54]

§5 [29],[34],[37]

§6 [7],[27],[40],[47],[48]

§7 [4], [32],[53]

§8 [8],[11],[16],[23],[44],[53]

LIST OF SYMBOLS

LIST OF SPACES

INDEX